Praise for **FOOLS RULE**

"As Marsden's book amply demonstrates, a new course on address-ing climate change is essential if we're going to save more human lives in the future." *The Georgia Straight*

"The more I read . . . I realized that William Marsden had found the right people to talk to, had identified who the players were—and also what the story was. And then it hit me: This guy really knows how to tell a story!" Steven Guilbeault, *The Gazette*

"With lively prose and a healthy serving of sarcasm and welcome indignation . . . Marsden . . . brings levity to this . . . analysis of the failure of the Copenhagen and Cancun climate summits." *Publishers Weekly*

Praise for **STUPID TO THE LAST DROP**
Winner of the National Business Book Award

"A must-read for every Canadian and for everyone around the world that idealizes and mythologizes Canada as a bastion of envi-ronmental stewardship." *GreenMuze*

"[Marsden brings] a fresh pair of discerning eyes to an unusual series of nation-changing events. He confidently reports how an entire province is destroying itself, and then asks why no one in Canada seems to care." *The Globe and Mail*

"A gripping and horrifying account of how the province of Alberta and the U.S. are ripping up tens of thousands of square kilometres of vital natural habitat to extract bitumen from the 'oil sands' in one of the most murderously polluting processes available to human beings." *New Statesman*

Also by William Marsden

Stupid to the Last Drop: How Alberta Is Bringing Environmental Armageddon to Canada (and Doesn't Seem to Care)

Angels of Death: Inside the Bikers' Empire of Crime
(with Julian Sher)

The Road to Hell: How the Biker Gangs Are Conquering Canada
(with Julian Sher)

FOOLS RULE

INSIDE THE FAILED POLITICS OF
CLIMATE CHANGE

WILLIAM MARSDEN

VINTAGE CANADA

Published in Canada by Vintage Canada, a division of Random House of
Canada Limited, Toronto, in 2012. Originally published in hardcover in
Canada by Alfred A. Knopf Canada, a division of Random House of Canada
Limited, in 2011. Distributed by Random House of Canada Limited.

Vintage Canada with colophon is a registered trademark.

www.randomhouse.ca

All photographs by William Marsden.

Library and Archives Canada Cataloguing in Publication

Marsden, William

Fools rule : inside the failed politics of climate change / William Marsden.

Includes bibliographical references and index.

ISBN 978-0-307-39825-3

1. Climatic changes—Government policy—International cooperation.
2. Climatic changes—Political aspects. I. Title. II. Title: Inside the failed
politics of climate change.

QC903.M37 2012   363.738'74526   C2011-903142-6

Text and cover design: Jennifer Lum
Cover image: JoannaBlu/Getty Images
Map: Erin Cooper

Printed on paper that contains FSC certified 100% post-consumer fibre

Printed and bound in the United States of America

2 4 6 8 9 7 5 3 1

To my mother, who looked for the best in everyone;
to my father, who disagreed

"We are blind ... we pervert reason when we humiliate life. Human dignity is insulted every day by the powerful of our world; the universal lie has replaced the plural truths; man stopped respecting himself when he lost the respect due to his fellow-creatures."
—*José Saramago*

"I think not saving energy is insanity. It is insanity from every point of view. It is insanity from economics; it is insanity from the prognostication of what is going to occur. It is so obvious, and it is so easy to do."
—Dr. Digby McLaren, president of the Royal Society of Canada and former director of the Canadian Geological Survey, testifying before a Canadian Parliamentary Committee on climate change, April 1990.

The ultimate objective of this Convention and any related legal instruments that the Conference of the Parties may adopt is to achieve ... stabilization of greenhouse gas concentrations in the atmosphere at a level that would prevent dangerous anthropogenic interference with the climate system. Such a level should be achieved within a time frame sufficient to allow ecosystems to adapt naturally to climate change, to ensure that food production is not threatened and to enable economic development to proceed in a sustainable manner.
—United Nations Framework Convention on Climate Change, 1992.

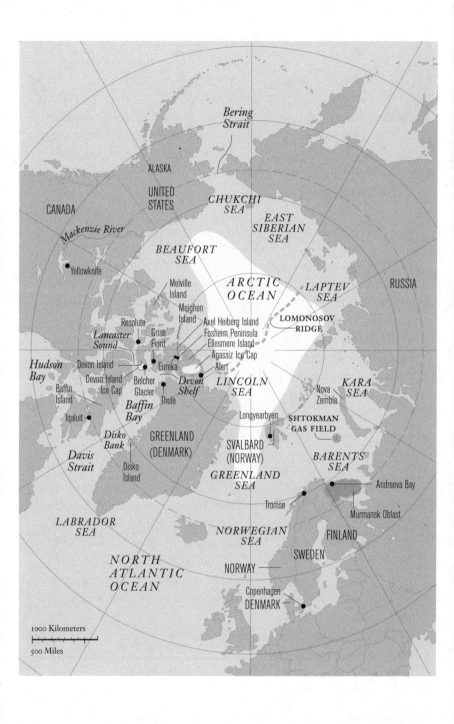

TABLE OF CONTENTS

TO THE COUNTRY FAIR

ARNE BANG MIKKELSEN WAS A HAPPY MAN. AND WHY NOT? THE convention had gone as planned. His logistics were flawless.

During the two weeks from December 4 to 18, 2009, when world leaders met in Copenhagen and spectacularly failed to produce a global agreement on climate change, Arne found success in feeding and watering them. The enormous food production system that mankind had been perfecting over the last eight thousand years—in the process conquering nature and altering normal climatic cycles—had worked. As chief executive of the huge hangar-like Bella Conference Center where the United Nations Framework Convention on Climate Change (UNFCCC) conference was held, he was "really proud," he said, that during the thirteen-day event the appetites of 45,000 people had been well served, the multitude having consumed three hundred tons of food including fish, poultry, beef, vegetables, fresh fruit and Danish hot dogs (*pølse*); 14,779 cakes (mostly apple strudel and chocolate squares); 350,000 glasses of water and 250,000 cups of coffee, plus thousands of bottles of beer and wine.

The only glitch was the long lineups into the convention itself, caused by a congested security system that forced some delegates to wait up to five hours in the cold of a Danish December before gaining entry. "The UN has apologized for this and has taken on the full responsibility," Arne said. Nothing was gonna stick to Arne. From the Danish organizers' point of view, the long queues were the only practical thing that did not function. "It has created respect throughout the world," they said after the conference wrapped up and the world leaders and delegates had, as Greenpeace put it, fled the crime scene.

Arne's finest hour, however, was not to be found in the simple fact of having fed so many delegates. As he stated in his final communiqué after the conference, it was the "record-time" assembly and furnishing of thirty-eight private meeting rooms, which the Americans and Chinese had ordered up with only three days left in the negotiations, that really showed his troops at their best.

Deep within the cavernous halls of Arne's Bella Center, where 192 nations struggled to quite possibly remake the world, it was in the seclusion of these rooms that a select group of world leaders leapfrogged the whole process and created what they called the Copenhagen Accord. Then they quickly saddled up their private jets and headed home to nations where the poor are clamoring for their fair share of the world's wealth or, in the case of President Obama, into a violent Washington snowstorm where the clamoring comes from a moneyed elite of "legal persons" with names like Goldman Sachs, Exxon, Chevron and Koch—the pillars of America's corporate democracy.

These private meeting rooms were where the Western powers attempted crudely and very publicly to bribe a defiant developing world into submission; where they tried but failed to sideline China and in the process reinforced the Communist country's overwhelming influence in the Third World; and where, as many scientists would conclude, the world's climate systems inched much closer to collapse.

Soon after they had gone, Arne's army swept away the evidence of failure in order to host another event in the annual rotation of fashion, holiday and car fairs. Within a week, the crime scene had been cleansed, erased. "Back to normal," Arne said. As if nothing had ever happened.

I followed the climate talks from 2009 to 2011, including the meetings in Bonn, Bangkok and Barcelona that led up to the Christmas pantomime in Copenhagen and then, one year later, the sun-splashed conference in the paradise of Cancún.

Initially, I was a parachutist landing amid a conversation carried on in an unfamiliar coded English. Words such as "Lu-Lu-CFs" (meaning Land Use, Land Use Change and Forestry), "Napas" (National Adaptation Programs of Action), "Redd" (Reducing Emissions from Deforestation in Developing Countries) or, my favorite, "BINGOs" (Business and Industry Non-government Organizations) were bandied about with the easy fluency of the insider. So arcane were these negotiations that I had to go to school in the language. Indeed, the United Nations supplied such a training for neophytes like me.

What was important, of course, was not so much the army of acronyms but the history behind them, something most delegates had long since forgotten. What had brought them to these meetings in the first place?

The answer was science. Relegated to trade show status, it had become a commodity you could take or leave depending on your needs. My journey through the science of climate change—particularly my trek over the Arctic glaciers to study their primal warnings—revealed the utter desperation of scientists as they pile proof upon proof only to see it disappear into the smoke of denial or crash against the excuse of political and economic expediency.

Science presents us with an assessment of risk. It tells us that climate change is the "defining challenge of our times," as UNFCCC executive secretary Christiana Figueres put it in the months leading up to the 2010 meeting in Cancún. "What is at stake here is none other than the long-term sustainable future of humanity . . . The milestone science has set . . . requires nothing less than an energy revolution both in production and in consumption." To achieve this, she said, nations need to grasp "the politically possible at every step."

Canada, which exhibits one of the more extreme cases of national cognitive dissonance, has turned back the clock on its greenhouse gas commitments, cranking up its tar sands production and even expanding coal-fired power plants. But the country is not unique. Australia, China, India and Brazil are all eagerly expanding their carbon footprints. These negotiations involve thousands of conflicting economic, social and political interests across individual, local, national and international levels that have so far defied a solution as each country marches along according to its greed.

Perhaps this is because the rich industrialized West is actively in denial as to what the stakes are. We act as if these negotiations are about politics as usual—Figueres's politics of the possible. Or, as Jonathan Pershing, the tall, self-assured American negotiator, told me: "The politics of the negotiations does not speak in any way to what *has* to be done." The science is overwhelming and frightening. But the reality is that the pace of political progress is a question only of achieving "milestones." Pershing is a scientist with a doctorate in geology and geophysics and an expertise in petroleum geology. He had previously worked as a climate change negotiator in the Clinton administration and also served as an author of the International Panel on Climate Change Fourth Assessment Report. So he should know better. Yet he sticks to the political mantra. While the politics is regrettable, he says, that is the way things are. The possible is always what's at issue.

Nations may find meaning in the politically possible, but climate change does not. It is a rising sea, a tsunami, an earthquake, a hurricane, a flood, a drought that sweeps away society's backup plans. It is a reminder that the way we live is not at all grounded in nature. The gap between what the science is asking us to do and what most people are willing to accept—what they claim is "possible"—gives you vertigo. "When you are at the table and you are negotiating a bit more tons or a bit less, it's insignificant compared to what you would need to do if you believe all these scientists," Canada's former environment minister Stéphane Dion told me.

Yet whether governments such as Canada's believe in the dangers of climate change hardly matters. What's important to them is economic stability so they can maintain social equilibrium and get re-elected. Laying down a carpet of deceit to calm social fears over global warming becomes a moral imperative.

How can you say you believe in the science and at the same time campaign against what the science proves is necessary to reduce the risk of runaway climate change? I asked Michael Martin, Canada's chief negotiator and ambassador for climate change, during an interview in Bonn in 2009.

"That's what these negotiations are for," he replied, adjusting his rimless glasses. "It's all about what is possible."

What about what is necessary?

"That's up for negotiation too."

If there is one inescapable issue in this entire affair, one question that encapsulates the whole sordid business of haggling over pollution, it is the matter of atmospheric space. How many more tons of greenhouse gases can we afford to put up there without causing catastrophic climate change, and which countries will get to emit them? Without a resolution of this issue, there may never be a deal.

The atmospheric space is the new frontier whose borders have gradually been defined over decades of scientific research. Like surveyors sent out to map new colonies and their potential to support human populations, scientists have charted the capacity of the atmosphere, the oceans and the forests—the earth's main reservoirs of greenhouse gases—to maintain a stable climate. They have discovered natural boundaries that they define in parts per million of greenhouse gases, mostly carbon dioxide. The normal carbon dioxide level in the atmosphere is 280 ppm. Our present level: 387 ppm, which puts us in the danger zone. We reach 450 ppm and we burn.

So far, our emissions have increased the mean temperature of the globe slightly less than one degree Celsius. But a global mean can be misleading. Arctic and equatorial temperatures have risen much more than that, and Canada's overall mean temperature has risen 1.3 degrees Celsius since 1945.[1] The issue at the climate talks is whether we should aim to limit the global rise to 2 degrees Celsius or 1.5 degrees. These are the numbers, by now familiar to most people who have followed the issue, that rattle around the halls and corridors of the negotiations. Rich nations argue for 2 degrees, the poor for 1.5. The motives are self-serving. The 2-degree figure gives the rich more elbow room to pollute; 1.5 degrees reduces the risk to poor countries who are absorbing the brunt of climate change and who have the fewest financial resources to adapt to its impacts.

The rich countries have historically and quite innocently claimed this atmospheric space for themselves. As they built their massive economies on the burning of fossil fuels, they dumped monumental amounts of CO_2 into the atmosphere unknowingly, at least in the beginning, reaching the limits of excess. The question now is whether the carbon space is full. If not, the rest of the world wants what's left. If the space is full, they want the rich countries to pull back drastically and surrender the carbon space to them. For the industrialized world, this would mean a major retreat in the face of the advancing emerging economies so that China and India and all the other countries that want our lifestyle can have their day. Alf Wills, a scientist and the chief negotiator for South Africa, said: "Until you can resolve . . . this linkage between ambition, global goal and equitable sharing of the remaining carbon space, there will be no agreement." Unfortunately, he said, developed countries have "no ambition" to go there.

And with reason. Because what they are negotiating, whether they like it or not, is a new world order. It's hardly something rich countries take lightly. A more egalitarian planet dictated by the carbon space reallotment means little to the Canadian tar sands worker or the American or Australian coal miner staring at unemployment. Such realignment might be far more morally and ethically defensible than our current predatory economic system, but it doesn't help these people. Nor does it answer the nervousness over tinkering with an economic system that has produced such enormous wealth and high standards of living in a matter of a few generations, if only for a relatively small proportion of the world's population. What would help is a willingness on the part of governments to face up to the realities of our time by preparing for a post–fossil fuel future, devoting massive resources towards technological development as well as harnessing the proven ability of society to change when change is needed. But this is not happening. Instead, we march ahead in total denial, licking our lips at the fossil fuel reservoirs buried in the melting Arctic and hoping the invisible hand of capitalism will save us. So far we have not even seen its fingertips.

If the West accepted a carbon space allotment, it would amount to a recognition of the enormous inequality that exists between rich and poor countries. It would constitute a voluntary retreat from economic dominance and signal a readiness to redistribute wealth. In the absence of a technology breakthrough which would replace fossil fuels with an energy system that can meet the ever-increasing demands of society and business, this is—for the next few decades at least—a zero-sum game. One country's loss is another's gain.

There are those who think otherwise. They believe that if industrialized nations greatly reduce their consumption and build clean-energy systems with the urgency that characterized the massive production scales of the Second World War, a quick transformation to 100 percent renewable energy is possible and everybody wins. But it's probably too late for that now. The size of the necessary reduction in emissions has become too big and the time frame is too narrow. In a world of limited atmospheric space where carbon is king and the best you can offer to replace it is sunshine and a windmill, zero-sum is the only outcome. For many poor countries already suffering under the strain of climate change, if the rich countries have to pull back, well, tough; there appears to be no other option, at least for the short term.

But the rich countries argue that if their economies suffer, everyone suffers; that any let-up in the pursuit of wealth will bring the global economy down on our heads like a house of cards, in which case there will be chaos. They deny the possibility of an orderly retreat. In lieu of any surrender of the atmosphere, they offer the climate change equivalent of sub-prime mortgages: a bundle of cash and technology transfer promises of dubious value to help poor and developing countries convert to clean energy and mitigate the effects of climate change. In return, the rich countries get to forge ahead with business as usual and the time-honored practice of screwing the weak.

Climate change negotiations have a unique political dynamic. Power at these negotiations does not derive simply from the size of your economy; it comes out of a chimney stack or an exhaust pipe. The more you emit, the more you can bring to the table. One of the sad realities in the struggle to meet the challenges of climate change is that the countries that pollute the most—the rich countries—hold all the cards. Within this group are the elite polluters: the United States, the European Union and China. They are the ones who have chips to deal, and so they rule the game. Countries such as Canada stand on the sidelines cheering for Team Industry. The rest of the world simply has the moral high ground, and rare is the historical moment when that has carried much weight. It is an undeniable fact that the countries who are the worst affected by climate change are too often poor countries who didn't cause the problem in the first place. There is, however, one leveler: climate change itself. Eventually, no country can escape that reality.

There is no end to the ironies created by climate change. The most powerful of these is the rising importance of the once-forgotten Arctic. The lure of great wealth plus control of new shipping lanes that could dominate commercial transportation in the northern hemisphere has the Arctic countries dreaming of a new world order run by them. Here too lies the new politics of climate change. Canada, Russia, Norway, the United States and Denmark are all rubbing their hands waiting impatiently for the big melt to release its spoils of minerals, oil and gas. Meanwhile, central Africa endures a rotation of unusual droughts and flooding depending on the time of year, but too often at the wrong time for crop planting. So populations scatter or die. Winners and losers face off at the climate change talks. The danger is the ultimate destruction of the global commonwealth.

Deniers, who constitute only a handful of ultra-conservative commentators, some political scientists and a sprinkling of scientists with dubious climate change credentials, keep repackaging debunked material and badgering legitimate scientists with irrelevant concerns. They speak to an audience of oddballs, happily angry at the world. Witness these messages sent by deniers to prominent climate scientists after the stolen emails from the climate research center at the University of East Anglia were made public:[2]

you, sir, are a nazi. go gargle razor blades, you fucking bastard!!!!!!!!

You are a fucking douchebag. You pathetic fucking Phony. I hope there is an earthquake right under your fucking house and swallows you into hell.

As a Lying worthless AGW [anthropogenic global warming] scammer, isn't it time you resigned and swam back to New Zealand. As a US tax-payer I want a fucking refund of all the wages you have fraudulently collected you asshole. Same goes for Jim THE FUCKING RAT Hansen [the NASA climate scientist]. Considering the state of our economy, maybe the public should begin the collection process.

We live in a public cyberspace where the scientific opinions of an oil company executive, a politician or a television commentator carry as much weight for the general public as the scientific knowledge of a geophysicist. Psychologists theorize that ultimately what drives many deniers is the obsessive need to be the smartest guy in the room even when they have no idea what they're talking about. The deniers can't dance, but they are convinced they're Fred Astaire. It's a world where the only important question is this: Why do people believe things that are patently false?

The answer may be buried in what the American novelist Upton Sinclair once said: "It is difficult to get a man to understand something when his salary depends on him not understanding it." He was talking about the difficulties of persuading people to vote for him as governor of California when they knew it would mean higher taxes. The colossal implications of climate change run far deeper than that; they challenge our view of who we are. Sinclair's observation would today go something like this: It is difficult to get a man to understand something when his whole self-image depends on him not understanding it.

There is no doubt that the denier chorus has cut a broad path through the evolution of climate politics. Copenhagen may have demonstrated that climate change has been elevated to the offices of presidents and prime ministers the world over, but that has not changed the pattern of its political progress, which has been one of plunging forward, stalling, retreating and then, well, we don't know. Denialism is both a useful idiot to vested interests and a power unto itself. The claim that deniers keep scientists honest is invalidated by the fact that they do it in such a dishonest way.

It would be great if the deniers were right. But that deafening drumbeat of science is unrelenting. It declares the fragility of our situation with greater certainty by the day. As we shall see, the task of tracking key climate indicators such as the movement of glaciers, the warming of oceans, the disappearance of fresh water and the extinction of species, is Herculean, often dangerous and terribly complex. But it proceeds unabated despite the many governments and corporate interests that seek to impede its progress.

When the United Nations created the International Panel on Climate Change to evaluate and report on the evolution of climate science, scientists thought we could fix the problem simply by reducing GHG (greenhouse gas) emissions to their pre-1990 levels and planting more trees. More than twenty years later, as we continue to increase our emissions, scientists have become resigned to a warmer planet and the question now is: How do we adapt? Proof of our growing desperation is that the efforts to geo-engineer the climate are no longer science fiction. Building giant mirrors in space to reflect the sun away from the earth is not garage talk, it's an agenda item in some scientific circles.

As delegates to the various UN conferences haggle over money and percentage points on CO_2 reductions, scientists meet in workshops where they disclose their most recent findings. In all cases, they show the frightening acceleration and mounting destructive capacity of climate change. Scientists and diplomats may find themselves in the same building, but they might just as well be on different planets—one that relies on the accumulation and assessment of hard facts, the other defined by the demands of industrial, mercantile and political interests. The science has galvanized the world into negotiations but not global action. The gap between science and political action continues to widen.

The challenge is rendered more complex by the fact that many countries have made little or no effort to educate civil society about the science of and dangers posed by climate change. In countries such as Canada, this has been purposeful. Canada, which was once a leader on the climate change political stage, is now led by right-wing conservatives from the tar sands economy of Alberta for whom unbridled capitalism is a religious edict. At best, they find climate change an annoyance. At worst, they refuse to believe it. Prime Minister Stephen Harper, who numbers some of Canada's most aggressive climate deniers among his closest friends,[3] has cut off funding to climate scientists and striven to muzzle scientists in the employ of the government, many of whom are leaders in their fields. Oil money has transformed the country. But Canada is not alone. Norway, Russia, Australia, the United States and, most recently, Greenland—all have succumbed to the continued lure of fossil fuels.

Few citizens are in a position to appreciate the enormous power wielded by corporate interests in countries such as the United States, where men like Senator James Mountain (Jim) Inhofe thrive in the world of the untrue. Inhofe is the senior senator from America's fifth-largest oil producer, Oklahoma, and the ranking Republican on the Senate Commitee on Environment and Public Works. The public record shows his campaign financing comes almost exclusively from the fossil fuel industry, with regular contributions from America's largest private oil refining company, Koch Industries, a major campaigner against caps on carbon emissions.

Inhofe arrived in Copenhagen on the last day of negotiations dressed in his signature black snakeskin cowboy boots, his face and hair ready for television, and leading his "truth squad" of one—himself. Unable to organize meetings with delegates, he corralled a herd of reporters around a staircase at the media center to tell them that climate change is a hoax started by the United Nations and the "Hollywood elites." But his main message amounted to what he claimed was a reality check: "Most of you are on the far left side, so listen closely. Nothing binding will come out of here in my opinion, and if it does, it will be rejected by the American people." Then the senator from Oklahoma, where fifteen coal-fired power plants annually produce about 36.1 million metric tons of carbon dioxide—a per capita production that is 50 percent above the U.S. average[4]—hopped back on a plane and flew home. Time spent in Copenhagen: about three hours. The modern version of hit and run. A farce? Yes. But a prescient one. The American government has not passed a climate change bill, and the likelihood that it ever will is remote. The vested interests opposing it are too powerful. In the United States Senate, it is far too easy to stop things. All climate change negotiators know that. Their trust in the American government and its horse-trading politicians is wafer thin. "It's the irony of the world that the fate of the world is being considered by some senators in the U.S. Congress," Ian Fry, the chief negotiator for the tiny Pacific state of Tuvalu, said. It's also an irony that in Copenhagen the only politician to speak the truth about America was the snake-skinned denier from Oklahoma.

Many people find hope in the "bottom-up approach." They target change in neighborhoods, towns and cities, where they believe small actions will lead to big results. They also believe the village is an incubator for political change. But global energy systems as well as global business interests are at play and, as you will see, without a global solution, the efforts of small communities alone won't cut it. Decisions made in Washington, Beijing, Mumbai, Paris, Berlin or London can sink the Tuvalus of this world.

The politics of climate change involves a cast of thousands. It is a clumsy tango of often halfhearted participants, many of whom have been out there so long it's looking like a dance marathon. They stagger about the floor inching forward, wavering and retreating. Two decades of global debate and still the emissions rise. Like a dithering, irresolute deal maker, we are allowing the opportunity to act to slip by. And then what? Modern society will likely demand that we do what Arne Bang Mikkelsen did: rejoice in our small organizational triumphs and prepare for the next food fair.

CHAPTER 1

THE MERRY-GO-ROUND

IN WHICH CANADA LEADS, THE WORLD FOLLOWS
AND BIG OIL SETS THEM ALL STRAIGHT

TIME WAS RUNNING OUT AS PIERRE PETTIGREW CHARGED AROUND
the convention center in Montréal looking for the Russians.

As he made his way along the concrete and glass corridors,
the Canadian foreign minister kept his cell phone to his ear, pray-
ing it wouldn't die on him. At the other end of the line was his
Russian counterpart, who had called from Moscow desperate to
speak to his delegation, whose members had inexplicably turned
off their phones.

It was December 10, 2005, and Montréal was in the grip of
its customary pre-Christmas freeze-up. Two weeks of difficult
climate talks had just about reached a consensus when the Russian
delegation suddenly refused to support an agreement on protocols
that would set the pace for negotiations in the years to come.
Then they had just as quickly shut off all their phones.

The whole thing was very odd, but Pettigrew couldn't afford
to dwell on that right now. The Russian foreign minister wanted
his rogue delegation to support the convention. Unless Pettigrew
could get the Russians to take his phone so their boss could set

them right, the entire convention would be lost, and quite possibly climate change talks would grind to a halt once and for all.

Pettigrew finally found the Russians in a far-off corner. He told them their boss was on the line and urgently wanted to speak to them.

They refused to take his phone.

Pettigrew insisted. *Take the phone.* Still the Russians refused.

On March 21, 1994, the United Nations Framework Convention on Climate Change (UNFCCC) came into force. Since then it has been signed by 194 countries. They have met in conferences in Berlin, Kyoto, Marrakech, Montréal, Nairobi, Bali, Copenhagen and Cancún, with many more stops in between. In the absence of an overarching technological solution to climate change, their goal is to negotiate protocols in the form of one or more international treaties that would lead to the "stabilization of greenhouse gas concentrations in the atmosphere at a level that would prevent dangerous anthropogenic interference with the climate system." That nutshell holds several important riders. The level "should be achieved within a time frame sufficient to allow ecosystems to adapt naturally to climate change, to ensure that food production is not threatened and to enable economic development to proceed in a sustainable manner." The element of urgency together with the commitment to continued economic growth makes the pursuit of a resolution the most challenging in the history of mankind.

The initial success was the Kyoto Protocol, in which thirty-seven industrialized nations collectively agreed to reduce their greenhouse gases by 5.2 percent of 1990 levels by the end of 2012. The United States, which at the time was the world's largest greenhouse gas emitter and is now second to China, did not sign the protocol. It has remained the only significant exception. And until the election of President Obama, the United States actively sought to undermine the negotiations by trying to persuade Australia, India and China to join it as a sort of satellite group outside the UN process.

The Kyoto Protocol was reinforced during the 2005 Montréal conference (known as COP 11, a reference to its standing as the eleventh Conference of the Parties to the UNFCCC) when the parties formally voted to implement the protocol and to start negotiations for long-term post-2012 emission reductions.

Since then, however, negotiations have floundered. The reasons are many and varied and not always evident in the smog of diplomacy. But in the end, most of the problem comes down to the basic human weaknesses of fear, greed and self-interest. All of which were thick on the ground at what the world hoped would be the decisive conference in Copenhagen. In setting that scene, we have first to look to Canada.

By Copenhagen, Canada had suffered a complete disintegration of its climate change policies. Its spirited commitment to ambitious action on global warming and world leadership towards an international treaty lay wretchedly broken. Staring gleefully at the wreckage on the floor were the perpetrators: a cabal of far-right Conservatives and their old guard oil company masters whose control of the Canadian economy and its government was now total.

These men—and they are all men—had slithered out of a province, Alberta, that had the second-largest store of oil in the world, a population of a mere three million and a remarkable talent for plunging into debt while the oil and gas companies sold off the province's resources, destroyed its ecosystems and made out like bandits. Now installed in the nation's capital, the oil oligarchy set out to transform Canada into a larger version of its triumph in the backwater of Alberta. It was no wonder, then, that by the time it arrived in Denmark, Canada had become a pariah to civil society and a useful idiot to its fellow eco-terrorists, including the United States and Australia.

Canada's struggle mirrored that of many industrialized countries. Its initial enthusiasm for creating a cleaner economy melted away in the heated debate over the costs of transforming energy sources: projected job losses and the claimed disruption of economic development. Particular importance was placed on Alberta's still-infant tar sands. Its reserves of at least 175 billion barrels of oil promised mesmerizing wealth. The fact that it would quickly become one of the dirtiest and most ecologically destructive industries in the history of mankind deserved no mention.

From 1988 to 1992, under Progressive Conservative prime minister Brian Mulroney, Canada had favored strong action that included setting reduction targets of up to 20 percent of 1990 levels. Mulroney aligned Canada with U.S. president George Bush, who in turn had been greatly influenced by British prime minister Margaret Thatcher, who believed climate change was mankind's greatest threat. But by the time of the Berlin conference in 1992, Canada and the United States had become more cautious. The two countries talked a good game, claiming they would freeze emissions at 1990 levels. But secretly they planned to align themselves with Saudi Arabia to disrupt the talks. The Saudis could always be counted on to protect their oil interests.

The federal government had been under pressure from the provinces, which jealously guarded their jurisdiction over natural resources and energy production, to do nothing that would jeopardize economic growth. Within the federal government there was a deep rift between the Department of Natural Resources, which advised against any targets that would disrupt such growth, and the much weaker environment department, which wanted strong action. A draft document prepared for the Canadian cabinet even mentioned using the Saudis as a proxy to upset the negotiations. "Sheila Copps [the Canadian environment minister] managed to have the reference to the Saudis removed, and the final mandate outlining Canada's position just said that Canada's interests were closely aligned with the United States," a senior Canadian climate change diplomat told me.

In Berlin, however, Canada's position changed after strong lobbying by the Germans and an about-face by the Americans. Canada agreed to sign the UN convention. Now it had to come up with emission reduction targets. After intense negotiations with the provinces, agreement was reached for a target limit of 3 percent below 1990 levels. This was the commitment the Canadians brought to Kyoto in 1997. The Americans, however, suddenly promised a reduction of 7 percent below 1990 levels. President Bill Clinton turned to Canada and pressured Prime Minister Jean Chrétien to match that level. Much to the fury of the provinces, Canada agreed to a reduction of 6 percent. Chrétien believed the difference of 3 percentage points could easily be made up through loopholes such as credits for safeguarding our forests or for building nuclear plants in China and elsewhere. Now it was a question of ratifying the treaty, which in Canada is a cabinet decision.

The Americans under their new president, George W. Bush, rejected Kyoto in 2001. Pressure came from both the American government and Canadian business communities for Canada to do the same. But Prime Minister Chrétien stayed the course. He was adamant that Canada support a critically important treaty that was in the public interest. Big business mustered its troops, flooding the media with dire warnings. Eight of the ten provinces, with oil-rich Alberta leading the way, expressed their opposition. With the Americans out of the deal, business claimed foreign investment would dry up, exports to Canada's most important market, the U.S.A., would suffer and 450,000 jobs would be lost. The oil patch announced that numerous projects would be put on hold if the treaty was ratified. Husky Energy and Canadian Natural Resources threatened to cancel billion-dollar tar sands projects or move parts of these operations to the United States. TrueNorth Oil Sands, owned by the billionaire Koch brothers from Wichita, Kansas, claimed it was reducing its capital budget and threatened to abandon the Foothills Oil Sands project, of which it owned 78 percent.[1] In a particularly nutty move, Alberta's Conservative government introduced legislation declaring carbon dioxide a natural resource and therefore the exclusive jurisdiction of the province. It also claimed it would lose $11 million every day following ratification, and even more if the treaty was implemented.

A lone voice among business leaders was Jason Myers, senior economist with the Canadian Manufacturers & Exporters association. He promoted a new vision: "Canada needs to take the issue of reducing greenhouse gas emissions seriously, and recognize that Canadians must invest heavily in new technologies and infrastructure to sustain our economy, environment and quality of life . . . The time to make these investments is now." But his words were lost in the racket coming from Alberta's oil patch.

Prime Minister Chrétien pressed on anyway. He bullied his divided cabinet by taking the vote to Parliament in the form of a resolution, declaring it a confidence vote and warning any Liberals who opposed him that he would not sign their nomination papers if the vote was lost and an election called. The result was unanimous Liberal support that won a Commons vote followed by cabinet approval of the treaty on December 17, 2002. "With this signature, we are doing the right thing for Canada, for the global environment and for future generations," Chrétien said.

Unique among the world's major fossil fuel producers, Canada committed itself to reduce its emissions by a staggering 240 metric megatons per year, or about 30 percent of 2003 emissions, by the 2012 deadline. That figure would continue to rise as Canada's emissions increased. "That Canada ratified Kyoto despite these challenges represents a triumph of ideas, in the form of commitments both to environmental sustainability and multilateralism, over economic interests and institutions," Kathryn Harrison, a political science professor at the University of British Columbia, said.[2]

The ratification, however, proved little more than words on a page as the battle moved to Canada's implementation plan. Opposition from the oil industry and Alberta was relentless, and stymied every attempt by the Liberals to reduce the nation's emissions. The federal government tried to implement an industry-wide plan that mixed real reductions with the purchase of carbon credits that would help finance development of clean-energy technology or the installation of clean energy in developing countries. Stéphane Dion, who at the time was minister of intergovernmental affairs, told me many companies were ready to accept this. The oil companies, however, were not among them.

The government argued that meeting the targets would shave less than half a percentage point off GDP growth over the next eight years, a figure that conforms to studies of the global economic effect of reducing emissions, all of which concluded it will be much cheaper to act sooner rather than later. The 2006 Stern Review for the British government on the economics of climate change, for example, concluded that "stabilization of greenhouse gas concentrations in the atmosphere is feasible and consistent with continued growth" and that "the benefits of strong, early action considerably outweigh the costs." Nations had a choice: they could spend 1 percent of GDP now or be saddled with as high as 14 percent or more in the future. Polls showed that most Canadians, despite the scare tactics from business, did not waver in their support for the treaty.[3]

Still, the fossil fuel companies, which produced half of the nation's emissions and whose profits were soaring well into the billions, deflected all pressure to reduce their emissions. Federal lobbying records show that they are the most active lobbyists on issues relating to climate change and the environment of any sector in Canada. Their voice is by far the loudest, and they have access to every major power center. They easily obtain face-to-face meetings with the minister and deputy minister of natural resources, with Canada's climate change negotiators and with the Prime Minister's Office.

Their persistence paid off. Soon after the Commons vote, in a letter dated December 15, 2002, to the oil and gas industry, the government promised that emission reductions for their sector would be set at "a level not more than 15 percent below projected business-as-usual levels for 2010." To meet their emission reduction targets, the government promised to allow them to buy carbon credits at no more than $15 a metric ton. If the price was over that amount, the government would pick up the difference. These targets were based on units of production (so-called intensity targets), with an industry-wide ceiling of 55 metric megatons. The companies were also promised generous tax write-offs on capital expenditures and equipment. Essentially, this would allow them to carry on with business as usual and the taxpayer would pay for any expenditures related to emission reductions. An industry whose annual net revenues from exports total between forty and fifty billion dollars[4] and whose projected contribution to the Canadian economy over the next twenty-five years is $3.6 trillion[5] apparently could not cough up two billion a year to reduce emissions or purchase credits that would help meet Canada's Kyoto targets. Yet even this concession on the part of the government was not enough.

The pressure continued until the summer of 2003, when the government finally relented. Chrétien himself drew up a letter to the Canadian Association of Petroleum Producers that set out eight policy principles designed to guide implementation of Canada's Kyoto commitments. Chrétien promised that Canada's emission reduction plan would not stall the normal growth of the oil industry, including the tar sands, and would not result in any job losses. The oil industry had got what it wanted: carte blanche to continue unfettered expansion.

The proof of failure lies in the fact that Canada's emissions, including those from forestry and land use, had risen 47.3 percent by 2008 from 1990 levels instead of decreasing 6 percent. Alberta, with less than 10 percent of the Canadian population and 19 percent of its GDP in 2008, produced 47 percent of the nation's greenhouse gases.[6]

Taxpayer subsidies and the lowest oil royalties in the world ensured that oil company profits remained high. Had Canada added, for example, a two-dollar green tax on every one of the 1.3 billion barrels of oil Canada produces each year, it would have gone a long way in helping to meet reduction targets and create jobs in the budding alternative energy sector. But the old guard of oil companies held firm. "It [Canada's green plan] is always designed in a way that you don't get in the way of oil sands development," one former senior Liberal adviser told me. "So that's the sacred cow and . . . you design it in such a way that it does not stop oil development and gas development and all of that. In particular, we have a democratic culture in North America of catering to the worst side of human beings . . . Europeans are democratic but they also have a culture of command governments. Governments are there to govern and governments don't mind making transformative decisions, long term and all of that. North Americans are just, well, energy pigs. And we cater to that." Even the automobile industry, whose vehicles create 25 percent of Canada's emissions, was exempt from emission reduction regulations. Despite majority public support for Kyoto, the government failed to stand firm. Canada had chosen the past over the future. It chose to subsidize oil rather than invest in the development of green technologies. The repercussions would fall on future generations.

Meanwhile, on the international front, Canada swung into action in 2004. Britain, France and Germany worried that the entire international process was slowly grinding to a halt because of the failure of Kyoto to gain global acceptance and to reduce global emissions. They pushed Canada to lead the way by hosting the COP 11 in Montréal, in December 2005. They believed that as an intensely industrialized country, second only to Australia in per capita greenhouse gas production,[7] Canada would serve as a beacon of commitment in the developed world. They also hoped it would have influence over its closest trading partner, the United States, as well as other oil-producing countries. "The challenge [of reducing emissions] was much bigger for Canada than for other countries," Stéphane Dion, who became president of COP 11, says. "It is, for instance, ten times more demanding for Canada to stop the growth of emissions than it is for France. If Canada can do it, anybody can do it."

Dion says the number one goal of COP 11 was "to save Kyoto." The country's experience with sophisticated multilateral diplomacy kicked in. Canada understood perfectly that the UNFCCC was a consensus process where trust between nations was critical. In order to gain that trust, Canadian diplomats spent the eleven months leading up to the conference traveling the world, listening to leaders in every region and discussing everybody's position. One of the diplomats involved recalls that most countries responded positively. The attitude was: "We need to be constructive, we need to do something." The diplomat adds: "Coming back . . . from our diplomatic tour, the very intractable players were India, Saudia Arabia and the U.S. Each one played hard, but they also knew they could hide behind the others."

For Canada, South Africa became a key ally. It had close historic relations with Brazil and India. The diplomat says Canada planned to use that relationship to isolate the United States if necessary. "The South Africans were more than keen to do their utmost to help us get to a deal, and we actually went three times to South Africa." Canada considered India a major impediment since, like China, it was desperate to build its economy and pull itself out of poverty. It didn't want restrictions on that growth. "As hardline as India can be, if it feels itself isolated, it will rally. So the South Africans kind of reassured us, 'Don't worry so much about India, we'll take care of it,'" he remembers.

The Canadians also learned that Saudi Arabia, which had been instrumental in slowing progress towards establishing ambitious global emission reduction targets, had become a pariah within an emerging power block at the UN, the G77 plus China. The G77 was formed in 1964 to represent the interests of seventy-seven poor and developing countries; it was an attempt to counterbalance the enormous economic and political power of the industrialized nations. Since then it has grown in number, although it prefers to keep its original name. Its membership is to a degree eclectic. Countries such as Saudi Arabia or the United Arab Emirates are richer than many industrialized nations. Its alliance with China, whose global search for resources has penetrated deep into Africa, Southeast Asia and parts of Latin America, has strengthened its voice. Western countries might be able to ignore Africa, much of Latin America and the Asia-Pacific, but they can't ignore China. With the emergence of both India and China as global economic forces, the G77 can be a formidable adversary if it sticks together. The Americans had for years been working through the Saudis and more recently through the Indians to hamper progress in the talks. Canada believed it had found a way to split that threesome by using the South Africans to lure away the Indians and the G77 to cast off the Saudis. "So what became clear in our mind was that we are going to get the Americans isolated too . . . and that we were able to isolate them, clearly isolate them," the diplomat says.

What has plagued negotiations all along is the "free-rider effect." This is where the gain from reducing emissions accrues for everyone but the cost is concentrated among the industrialized nations. This had become a major political hurdle. Kyoto had required that only industrialized countries reduce their emissions. The idea was, they caused the problem, they should fix it. But countries such as India and China had by 2005 emerged as mega-polluters. China would soon become the biggest greenhouse gas emitter in the world. It was building several coal-fired power plants every month. These would easily negate any emission reductions by industrialized countries.

To get the United States onside, Dion had to persuade the Chinese to accept targets. But given Canada's and the United States' historic record as major polluters, it would be difficult. Furthermore, China demanded major technology transfers before it would agree to targets. Dion said China also wanted money to help it finance projects to adapt to climate change. "They look you in the eyes and say, 'Are you asking us to sacrifice? We have hundreds of millions of human beings without electricity and our job is to bring electricity to them. Don't ask us to wait when you are coming in your big limousine and big jet and your emissions and I see in your plan that you are asking your citizens to decrease their emissions by one ton—the one ton challenge. Well, we are dreaming of the day when our citizens will emit at least one ton, because then we will be rich.'"

China by 2005 was already on its way to imposing domestic targets to deal with the local effects of its smog problem. But it refused to negotiate internationally monitored targets because it wanted the United States to agree to targets first, Dion says. But China did not surface as a major issue at COP 11 in Montréal in 2005 because Dion was determined to keep the focus precise: preserve Kyoto by getting the parties to agree to begin negotiations for post-2012 targets.

The Montréal conference got off to a good start. The 192 countries immediately voted to accept the underlying regulations of

Kyoto and agreed to begin the process of helping poor countries adapt to climate change. Yet by the end of the two-week meeting, Dion had not persuaded the delegates to agree to start negotiations for post-2012 targets. The main laggard was the Saudis, who were being used by the Americans to defeat the process and who were trying to create solid opposition throughout the developing and undeveloped world. "The other parties told me that they preferred no deal to a bad deal and if the Americans are creating a bad agreement, forget it," Dion says. "The Americans didn't want to do anything, but they didn't want to be blamed [for a failure]. Even though they were not a part of Kyoto, they were strong enough to kill it." Dion intended that the Americans should know that if they stopped negotiations for post-2012 targets, they would be held responsible for the eventual death of the Kyoto Protocol, mankind's only international climate change treaty. Although the Americans were not part of Kyoto, they were still part of the United Nations Framework Convention on Climate Change, which had created Kyoto. Therefore they still had standing and their superpower status gave them a powerful presence.

Two days before the end of the conference, Dion heard from his deputy that the American delegation was ready to leave. He immediately arranged to meet with U.S. lead delegate Paula Dobriansky, the under secretary of state for democracy and global affairs, who told him: "We have created a new group with Australia, China and India." The United States, it seemed, had successfully corralled the world's largest polluters into a bloc of opposition and they were all ready to walk out. At that point Dion thought the conference was dead and the Americans had won. They and their breakaway group had killed Kyoto. But Canadian diplomats, some of whom were of Arab, Indian and Chinese descent, began working the phones to contact opposition leaders in those countries, hoping they in turn could persuade their national leaders to break free of the Americans. It was Dion's belief that the national leaders of these countries did not fully grasp the seriousness of the situation.

Canada's homework paid off. The Saudis had constantly slowed the negotiations through procedural moves, largely at the urging of the Americans. They wanted to protect their oil industry and were glad to have the Americans silently at their side. But their big hope was to drag a concession out of the conference to compensate them for any loss of oil revenue as a result of Kyoto. They knew there was less than the slimmest of chances that they would ever win such a concession, but it was a useful ploy to bottle up negotiations and impede progress towards a clean-energy world. Canada's diplomats were aware of the American-Saudi connection; Canada itself had used the Saudis at past meetings. The diplomats also knew that many oil-producing countries in the Middle East did not support the obstructionist Saudi position. They believed that these countries would desert Saudi Arabia and leave it isolated. Diplomatic moves did just that. The G77, which was never comfortable with the Saudis, turned against them. In addition, a senior Canadian official, of Arab descent, had a word with the Saudis. In the words of one delegate, he told them: "The highest value of an Arab is our sense of hospitality. We have treated you well and given you all of the opportunity you need, so don't fuck with your host."

With the help of the South Africans, the Indian delegation also suddenly became more supportive, as did the Australians. Dion knew that the Chinese were reluctant to follow the American lead. They had supported Kyoto and had never been happy with America's rejection of the treaty. America was isolated. Yet it would not relent and kept up the pressure on its wavering allies.

The final nail in America's coffin was hammered in by former U.S. president Bill Clinton. Elizabeth May, who had known Clinton since childhood[8] and who at the time was director of the Sierra

Club of Canada, had invited Clinton the previous June to speak to the environment groups attending the convention. Before every UNFCCC conference, the environment groups jointly chart strategies and divide up responsibilities. Part of May's mandate had been to raise the profile of the Kyoto Protocol and the COP 11 in the U.S. media, which had largely ignored the event, believing that Kyoto was dead. May's view was that Clinton's outspoken support for Kyoto and for ambitious action on climate change would resonate throughout the U.S. media were he to speak in Montréal. In the middle of the two-week conference, May got confirmation that Clinton could come on the last Friday of the convention.

News of his coming initially rattled the Canadian organizers, who worried that it might be too provocative and anger the U.S. administration, thereby solidifying their opposition. After considerable discussion, however, they decided that, if handled delicately, this would be a good opportunity to embarrass the Americans into throwing their support behind the convention and the continuation of Kyoto. "This is not normally Canada's way," one Canadian delegate says. "We try to be allies with them. But we felt we were in good shape to isolate them. The Clinton speech was choreographed. You are not supposed to play politics and so it was billed as a surprise appearance."

Though the Canadians tried to keep the visit a secret until the last possible minute, the Americans quickly found out, and their reaction was swift and angry. One Canadian delegate quoted them as basically saying, "If you let Clinton in the building, we walk."

"This infuriated the U.S. delegation," Dion says. "And I thought, 'I'm dead. It's the end of the story.'"

May recalls, "They had been threatening to walk out for a couple of days at that point." She says she asked a senior Canadian delegate if she should cancel the invitation. "And he said, 'No no, the government of Canada would never ask civil society to uninvite someone just because of a threat from another government.'"

Canada did not want Clinton's speech to be a simple side event. Yet since he was in Montréal at the invitation of an environment group, he would not be able to speak to a plenary of the convention. So during a break in the plenary, the Canadians turned all the chairs around to face the back of the room, where there were no UN logos, nothing but a blue curtain. The hall was then reopened as a side event and all the delegates piled back in to hear Bill Clinton.

He spoke standing at a clear Plexiglas podium without notes. He talked of the science of climate change, the technology of clean-energy systems and the economic benefits they would bring. He spoke of the duty mankind has to future generations and of the need to unite in our efforts to reduce greenhouse gases. The speech was a success not only in what he said, but also in what he didn't say. He did not blame or even mention the Bush administration or say anything that would embarrass it. He didn't have to; the implication was clear. The speech was a harsh indictment of American intransigence. He ended by saying: "So, again, my plea is for us all to get together, let's try to go forward together. And if you can't agree on a target, agree on a set of projects so everybody has something to do when they get up in the morning. This is a terrible thing to paralyze ourselves, and give people an excuse, and let anybody off the hook from doing something. Let's find a way to walk away from here and walk into the future together, so that we all have something that will give our grandchildren this planet in a more prosperous and more humane way."

Clinton is a powerful orator with a way of reaching out to every individual in the audience and drawing them to his side. He got a standing ovation. When the plenary restarted, the Americans, on orders from the White House, had changed their position and agreed to support negotiations for a second Kyoto commitment period. A consensus, it seemed, could now be reached.

Then suddenly the Russian delegation entered the ring. Until now taciturn spectators, they abruptly refused to support a second commitment period unless the convention met their demand that the Kyoto Protocol be amended to permit former Soviet republics such as Belarus and Ukraine to sign on as Annex B countries (the category for industrialized countries). Participation as Annex B countries would allow them to benefit from the carbon markets, which could mean billions of dollars' worth of credits for these low carbon emitters. Russia had been quietly trying to negotiate their entry throughout the conference, but no one had taken the request seriously. Entry into Annex B is a highly complicated process and the conference was not mandated to make such a decision.

The Canadians were stunned that the Russians would use this justification to hold the entire process hostage. Rumors flew that the Americans had put them up to it. The Canadians doubted that Moscow supported the new Russian position.

South Africa's lead negotiator, Alf Wills, and Canadian diplomat David Drake were sent off to negotiate with Russia's head of delegation, Alexander Bedritsky, a meteorologist who had claimed that climate change would be good for Russia because it would warm the north and who had been skeptical about Russia signing Kyoto because the economic benefits "were illusory."[9] Wills says they offered him a written promise that the issue would be tackled in the lead-up to and at the next conference in Kenya. But Russia held firm and there seemed no solution. It was just before dawn on Saturday morning and the convention was exhausted. The interpreters, who were being paid by NATO, made threatening noises about leaving because they had been working unauthorized overtime for twelve hours and didn't know if they were going to be paid.

Meanwhile, Canadian prime minister Paul Martin called Moscow and talked to Russian president Vladimir Putin; Canada's foreign minister, Pierre Pettigrew, contacted the Russian foreign minister. The latter told Pettigrew that the Russian environment minister was not acting according to Russian policy. He promised to call Bedritsky and solve the problem. But nothing happened. Then several hours later, the Russian foreign minister called Pettigrew back and told him he couldn't get through to his delegation—they had turned off their phones. Can you please take your phone and find my delegation? the minister asked. So Pettigrew searched the convention center for the Russians, while trying to keep his phone open. He finally found them in one of the building's many corridors. "I have your foreign minister on the phone," he said, and offered them his cell phone. At first they didn't want to take the call.

Dion understood their reluctance. "I wouldn't have accepted a phone call from my minister on a phone offered by a foreign country," he says.

But Pettigrew insisted they take the phone; the minister could be heard bellowing over the line. Finally the Russians agreed to speak to him. When the call was over, they marched off and notified the convention that Russia no longer had any objections.

At the end of an all-night session, the conference voted to begin the next round of negotiations for post-2012 targets. They also created a second track, as yet undefined, intended to bring the whole world into an agreement for long-term emission reductions by all countries, developed and undeveloped, under the United Nations Framework Convention on Climate Change, which the Americans had approved in 1992. Both of the traditional holdouts, the Chinese and the Indians, supported this move. The task for future meetings would be to define that track.

Montréal was a turning point if only because it put the negotiations back on track. A fatalistic fatigue had set in. Now there was some light. The Americans, Australians, Chinese and Indians had locked themselves into becoming full players in future negotiations, the parties were re-energized and the way forward seemed clear.

———

The next meeting was scheduled for a year later in Nairobi. Canada wanted to maintain the momentum. It put aside $250 million and a staff of more than one hundred civil servants to help the Kenyans in the run-up to the next meeting and to continue multilateral discussions over the next five years that would lead to a new long-term agreement at what would become the Copenhagen conference. First Canada intended to begin an "all-out diplomatic effort" to help the Kenyans keep the ball rolling. The government established a special budget to help finance the Kenyans. If the world's fifth-largest oil-producing nation was ready to demonstrate a leadership role in the pursuit of a lasting treaty on climate change, that alone served as a powerful message of commitment and trust.

But one month later, the Conservatives beat the Liberals in a federal election and formed a minority government. The oil industry was now firmly in control in Ottawa.

The new government acted quickly to destroy what the Liberals had created. They disbanded Canada's international climate team and canceled the budget. There would be no diplomatic effort. The knowledge gained prior to and during the Montréal conference dissipated. The Kenyans and whoever came after them were on their own. The Liberals' Green Plan project to begin the transformation of the nation's energy grid to clean technology was shelved. Finally, the Conservatives erased a wealth of data the previous government had been compiling since 1998 relating to the energy needs of Canadian municipalities, transportation, agriculture and industry. What had been a story of ideas for a transformative future dissolved into the reactionary thinking of a fading fossil-fuel age. Canada had become what it had fought almost a century and a half to avoid: small change in the American purse. It simply abandoned the field and became the silent partner of the United States. The consequences are still being felt in the distrust that has since corrupted the progress of the climate talks.

———

One country viewed Montréal's success as marginal at best. Early in the 1970s, when OPEC suddenly jacked up oil prices, Denmark's economy was almost paralyzed. Danish roadways were emptied of cars and trucks. Bicycles took their place. The Scandinavian country spent the next thirty years transforming itself from a country that imported more than 95 percent of its energy from the Middle East into a renewable-energy powerhouse. In the process, it became a leader in the design, production and use of green technology. While the Danes still use coal as a backup to wind turbines, natural gas and biomass—they refused to go nuclear—they are laying the plans for a fossil fuel–free future. Since 1980, the Danish economy has grown 78 percent while lowering its greenhouse gas emissions and stabilizing its energy consumption. For the Danes, their success was proof that a country can substantially reduce its use of dirty fuels while improving its standard of living and national wealth. By 2005, one Danish politician was eager to take the Danish model global.

CHAPTER 2

DENMARK'S HUBRIS

IN WHICH THE DANES WARN OF PLANETARY SHUTDOWN AND BUILD A FIRE UNDER WORLD LEADERS

CONNIE HEDEGAARD HAD BEEN IN AND OUT OF POLITICS SINCE 1984, when she became the youngest member ever elected to the Danish parliament at age twenty-four. She quit in 1990 to become a newspaper journalist, later moving into television, where she was the presenter for Denmark's most popular current affairs program. By 2005 she was back in parliament with a raised profile and a cabinet position as environment minister. She says she believed the time had come for her to make her mark.

Her party of long standing was the Conservative Party, but it was not at all conservative in the North American sense. "It means preserving the environment and handing over to future generations a safe world and the same opportunities we have today," she tells me in an interview. The Conservatives were only one of nine parties in Danish parliament. They ranked fifth from the top in the number of seats, but their close alliance with the Liberals, a center-right, pro-business party, gave them several cabinet positions, including the environment portfolio for Hedegaard. Her Liberal Party bosses, however, did not always agree with her conservation enthusiasms.

Her prime minister was a fan of one of Denmark's most notorious climate change deniers, a political scientist named Bjørn Lomborg whose 2001 book, *The Skeptical Environmentalist*, caused a furor among Danish climate scientists. Despite—or perhaps because of—his lack of a scientific background, Lomborg's book dismissed climate science out of hand. But the Liberals liked him so much that they made him head of the Environmental Assessment Institute, which the government created in 2002 to evaluate the costs and benefits of environmental action. Lomborg may have been neither an economist nor an accountant, but he immediately argued that pumping money into climate change was a waste of capital. By 2010, he had changed his tune. He claimed that he now believed in anthropogenic climate change but thought it was too late to do anything about it other than spend money on technology development and hope for the best. That a government claiming to be serious about climate change would sympathize with Lomborg made a clash with Hedegaard inevitable.

Hedegaard was looking for a world stage. Former adviser Per Meilstrup recalls long discussions with her about what would be the next big issue where Denmark could make its mark.[1] "It wasn't hard to identify global warming as a main and overriding challenge," he says. "She thought that it would be logical for a country that had actually tried to deal with these issues to show leadership."

The question was how to grab that stage. She decided she would hold her own international climate talks, and organized what became known as the Greenland Dialogue. First held in August 2005 in Ilulissat, Greenland, it was an informal closed-door meeting of ministers from twenty-two countries including Canada, representing most of the major economies as well as some underdeveloped and developing countries. A noticeable absentee was China; Canada too would pull out of future meetings. The mandate was to discuss how to accelerate the negotiations towards a final global agreement. Ten main points surfaced. Among them was the fundamental need to raise global awareness of the dangers of climate change and to broaden the scope of the political negotiations. Essentially, the world had to figure out how to divert the estimated US$16 trillion it would spend by 2030 on the energy sector towards a clean and sustainable future.

Montréal was Hedegaard's first United Nations climate change conference. She was not impressed. It may have ended in success, but that success, in her eyes, was insignificant when measured against the task. It was as if the world had not woken up to the scale of the issue it faced. At one point she sat down with her chief adviser in the lobby of the Montréal convention center and they began discussing what Denmark might do to lead the world to a final agreement on climate change. The more they talked, the more enthusiastic she became. She believed the talks had floundered because they were not taken seriously at the highest levels of government. "We had to bring it onto the agenda of political leaders," she recalls.

World leaders had been involved in the distant past. Margaret Thatcher had sparked negotiations in the late 1980s. Bill Clinton and Al Gore had pushed them forward in the mid-1990s. But those politicians were gone. Gore was now a lone salesman peddling his climate change documentary, *An Inconvenient Truth*, on the fringes of a doubtful world caught up in exciting economic growth, the emergence of Asian economies, the terrorist attacks of 9/11 and the Afghan and Iraqi wars. Canada was ripping up its boreal forest to get at the tar sands; Australia was shipping massive amounts of coal to China, which was constructing coal-fired power plants at a frenzied pace at home while razing Africa's jungles and carting away exotic lumber and hidden mineral wealth; Indonesia was sacrificing its rain forests for palm oil plantations; Brazil was feasting on the Amazon.

Against this backdrop of feverish economic action, climate change was just a spoiler. It wasn't any fun. Let a lowly environment minister handle it. Hedegaard disagreed. She intended to dump the climate change file back onto the desks of national leaders. Drag them back to the bargaining table. Make it the biggest, most pressing issue in the world. "It's about north-south, rich and poor," she says, recalling her excitement in 2005. "It's about financing. It's about energy access. When you have huge topics like that, how are you going to get that moving without having the backing of the heads of state."

When Hedegaard returned from the Montréal conference, she proposed to cabinet that Denmark apply for the presidency of the 2009 COP. Because the first commitment period under the Kyoto Protocol expired in 2012, she figured she could make this the decisive conference where the world would approve a new all-inclusive treaty that would finally unite all nations in the fight to stabilize the climate. The Danes would brand the world with a green label. And they'd sell a few wind turbines in the process.

After Nairobi in 2006, which failed to advance the agenda, Denmark got its wish. The United Nations awarded it the December 2009 conference known as COP 15. And the debacle began.

Canada had been a reluctant president. It accepted the job largely at the bidding of Britain, France and Germany, who believed that Canada's history as a trusted player in United Nations negotiations and its faint influence—oil—on the United States would advance the deal-making process. Canada's federation has been held together by consensus building and respect for the sovereignty of its provinces. It understands multilateral diplomacy perhaps better than anybody. It takes to the task of consensus building naturally because its very existence as a nation depends on it. At the Montréal COP in 2005, it understood the role it was about to play because that role is in its blood.

Denmark, by contrast, wanted to globalize a Danish vision. Its participation was more about Denmark imposing its view on the world stage than about seeking international consensus. If it had to leapfrog the UNFCCC process to get a treaty, then so be it. But as host of the conference, Denmark was supposed to be working for the 192 participant countries. It was there to facilitate the negotiations and help navigate the pitfalls. It was not there to lead, certainly not from the front. As president of the convention, Hedegaard was supposed to be a servant of the conference.

She was not unaware of the dangers of trying to impose Denmark's will on the convention, and initially she went through the motions of establishing trust. Denmark set up climate envoys in all the capitals that it believed would be critical to a successful outcome. Hedegaard says her goal was to solve the so-called prisoners dilemma. The process so far had seen rational self-interest triumph over group interest despite the fact that serving the interests of the whole group would be beneficial for all its members. As usual, nations were simply out for themselves. How do you get nations to cooperate for their own good? "We were spending a lot of energy on how to take care of this issue," she says. "We have seen on international climate negotiations for much too long that everything is depending on everything else and everybody is waiting for everybody else. Would there be any way that we could sort of break that?" One way was to try to get emerging nations and the United States to sit down at the same table. If the emerging nations—primarily China, India and Brazil—and the Americans could agree to accept responsibility for emission reductions, then all other nations would fall into line.

Out on the diplomatic circuit, Hedegaard and her climate envoys may have struck a consensus tone, but back home in Copenhagen the attitude was one of impatience. Prime Minister Anders Fogh Rasmussen, according to Meilstrup, "had little faith" in the UNFCCC process. He thought Denmark should be more forceful in the negotiations leading up to and during the Copenhagen meeting. Denmark had recently held the presidency of the European

Union during its expansion into Eastern Europe. That had involved hard, multilateral negotiations, but in the end the Danish-led EU had succeeded by imposing its will and banging out a final deal. The Prime Minister's Office thought the country should be just as forceful in the climate change negotiations. Hedegaard disagreed. While she and her prime minister were of the same mind in believing the UNFCCC had so far been a failure, Hedegaard realized Denmark needed to work as much as possible within the convention and that certain diplomatic niceties had to be observed. She wanted at least to give the appearance of consensus seeking. "[The expansion of the EU] was very, very different from hosting 192 countries in Copenhagen," Hedegaard says. "And that was one of the basic differences where we said, 'Listen, we have to respect that this has to be a party-driven process.'"

The internal confusion over its role gave Denmark the appearance of a split personality and led to struggles within the Danish bureaucracy and the cabinet that hampered the process. Yvo de Boer, executive director of the UNFCCC, later claimed that he often didn't know who was talking, Denmark #1 or Denmark #2—the consensus Denmark or the assertive Denmark.

In April 2009, eight months before Copenhagen, Prime Minister Anders Fogh Rasmussen resigned to take the job of secretary-general of NATO. His replacement, Lars Løkke Rasmussen, proved even more impatient with the UNFCCC process. Hedegaard was the official president of the COP and was in charge of a new climate and energy ministry that was tasked with organizing the conference. But the new prime minister chose to create a parallel climate bureau in his own office through which he could exert his influence, particularly with his fellow world leaders. Experienced Danish negotiators who cautioned Rasmussen against taking such a strong-headed approach were sidelined. Hedegaard soon discovered that Rasmussen was going behind her back in organizing one-on-one meetings with other governments.

Hedegaard's and Rasmussen's staffs were at constant loggerheads. A few months before the conference, Hedegaard threatened to resign, then changed her mind. But the bickering remained so petty that Denmark's chief climate negotiator, Thomas Becker, resigned after Rasmussen's office claimed he had violated ministerial regulations by spending too much money on the wining and dining of delegates at conventions in Bonn and Bangkok. "It wasn't as if he was lining his pockets or anything like that. Somebody who wanted him out got him out on these little things," says Zammit Cutajar, a former director of the UNFCCC and chairman of the negotiations for Long-term Cooperative Action.

By the time the curtain rose on COP 15, Denmark had already lost the plot.

Danish hubris wasn't the only problem that reduced the likelihood of success as Copenhagen approached. Two other elements came into play.

The hype surrounding the conference was enormous—and intentional. "Its timing was spun as the 'last chance to save the planet,'" Cutajar says. The Danes consciously sought to create a sense of extreme urgency, a now-or-never atmosphere, that would make the political price for not acting too high for any politician. For Hedegaard, this was simply conveying to the public the reality of climate change: that the price of inaction would be catastrophic for everyone. Two years earlier, in Bali, the parties had decided to set Copenhagen as the deadline for an agreement. "It was not just sort of an invention of Denmark," she says. But Denmark had upped the ante by using the deadline as an excuse to detour around the consensus process.

In the months leading up to Copenhagen, secret Danish texts were handed out to various select parties. This alarmed the developing world. Many countries suspected that the West was subverting the UN process by attempting to impose a weak agreement on the conference and shirk its responsibilities.

The goal of Copenhagen was twofold: agree to post-2012 targets for the Kyoto Protocol and at the same time negotiate a parallel treaty that would include the United States and would also impose long-term emission reduction targets on all countries. Furthermore, this agreement was supposed to create institutional structures and financial arrangements that would help primarily poor countries adapt to climate change and build green energy economies.

In Bangkok, open discussions about eliminating the Kyoto Protocol for the first time filled the hallways and gained legitimacy in press conferences. Industrialized nations were now publicly demanding that developing countries accept emission targets, particularly the emerging economies of China, Brazil and India. The West threatened to withdraw money for adaptation if they didn't.

This issue had surfaced before, in 2006 during the second Greenland Dialogue in South Africa. The meeting decided that the Kyoto Protocol and the Long-term Cooperative Action treaty would eventually have to merge into one convention; to have two global treaties seemed counterproductive, unwieldy and ultimately confusing. "While this does not appear feasible immediately, a bridge needs to be created between what is feasible and what is necessary," the twenty participants concluded. In Bangkok, the issue suddenly came out into the open, where it fostered more distrust. The Europeans, Australians and Canadians supported one treaty; China and many developing countries vigorously opposed it, arguing that they wanted to hold on to the only existing international agreement with the legal status to commit the industrialized world to emission reductions. The supporters of Kyoto noted that it had taken almost two decades to negotiate and approve it. To kill it now would likely mean waiting at least another ten years while a whole new treaty was negotiated, without any guarantee that it would be ratified by all 193 national governments.

With its first commitment period ending in 2012, the Kyoto Protocol was running out of time. Certainly, an end to Kyoto would be ideal for countries such as Canada, which had already thrown in the towel, admitting that it would never reach its target of 6 percent below 1990 levels. In fact, the country fully intended to allow its emissions to increase over the next five years at least. Its unrestrained tar sands industry was announcing a doubling in its projected greenhouse gas emissions over the next decade. (In 2009 alone, tar sands emissions increased 11 percent, and it remained the only industrial sector that had seen its greenhouse gases rise.[2]) Canada was eager to avoid financial penalties for non-compliance that could cost it billions of dollars. If countries failed to meet their obligations within one hundred days after the 2012 deadline, the Kyoto compliance committee could force them to purchase emission credits to make up for the shortfall. Non-compliant countries would also have to make up the difference in the form of deeper emission reductions and add another 30 percent to their targets as a penalty. In Canada's case, it would have to spend billions of dollars buying emission credits from countries that had met their targets and then could be barred from reselling those credits on the carbon market. So the price of non-compliance could amount to an astronomical penalty. Canada's solution was to try to kill the Kyoto Protocol by replacing it with something that would allow non-compliant nations to escape any penalties.

China, the world's dirtiest polluter on every possible level, and India, which was beginning to hold its own on that score, felt the most exposed and pushed back the hardest, demanding that Kyoto be retained. In the months leading up to Copenhagen, the tone was conspiratorial, acid and highly confrontational.

The Danes had invited the world to their doorstep, but now they found themselves staring at defeat before the guests had even arrived. The more desperate the talks grew, the more determined were the Danes to impose their will.

CHAPTER 3

TRAGEDY IN COPENHAGEN

AN ERRANT INVITATION, A BLUSTERING AMERICA AND THE EMERGENCE OF A NEW GLOBAL FORCE

Voices from Cyberspace:

Gelion:
The truth of the matter is that the West and the rich countries do not want to pay for the havoc that they instilled by unregulated global capitalism and trying to grow their economies by 3%+ a year.
1:43 PM April 12, 2010 from web

Bartelbe:
If the west says to the BASIC nations [Brazil, South Africa, India and China], we want a climate deal and if we don't get one we will block your access to our markets, you would get a deal. I doubt the green movement, with its links to the left on the political spectrum, would support this. They would go on about the West's pollution in the past, the immorality of the rich telling the poor not to do what the rich have already done. They would be morally in the right. Unfortunately, there is no such thing as moral green house gas. The pollution of [sic] the poor is just as damaging as the pollution from the rich.
2:45 PM April 12, 2010 from web

KnowThankYou:
We have to understand that our governments are no longer working for the people, they are working for the economies, and as long as this remains the case corporatism will endanger negotiations of this kind and others.
3:04 PM April 12, 2010 from web

ALF WILLS BEGAN COUNTING HEADS. IT WASN'T CONSCIOUS. IT was just one of those idle things people do as they wait for something to happen. Who's present. Who's not. It wasn't easy, either. The Copenhagen conference room was small and there was a lot of chaos as world leaders and their aides milled about. Instinctively, though, Wills sensed that something was wrong.

Before him were assembled the world's most powerful leaders: U.S. president Barack Obama, French president Nicolas Sarkozy, German chancellor Angela Merkel, British prime minister Gordon Brown and Japanese prime minister Yukio Hatoyama. Also in the room were a group of leaders from key developing countries such as Brazil and India, and Wills's own South Africa. The group of twenty-nine nations had gathered this Friday morning—the last day of the conference—to try to bang out a last-minute climate change agreement. They constituted the world's major polluters—except one. That's what had jarred Wills's thinking as his tired eyes scanned the room. Where was China? Where was Chinese premier Wen Jiabao, the leader of the world's biggest greenhouse gas polluter?

The evening before, at a state dinner, the Danes had made a final attempt at salvaging the negotiations, not to mention Danish pride. They proposed that a select group of leaders meet to discuss a three-page text—the latest in what had become a series of secret Danish texts—aimed at producing a political agreement that summarized common ground and that would be grandly called the Copenhagen Accord. Wills had received a phone call from his president that evening instructing him to meet through the night with the negotiators of this select group to review the proposal. It was a mad dash for the finish line.

The sense of urgency was palpable. Of the 192 countries attending the conference, 122 had sent either their head of state or their head of government. The world had never witnessed such a grouping outside the United Nations headquarters in New York. But almost from the start the conference had collapsed into what the U.S. secretary of state, Hillary Clinton, told President Obama was "the worst meeting I've been to since the eighth-grade student council."[1]

The Danes' last desperate attempt to come up with some kind of agreement began after that state dinner, at eleven-thirty Thursday evening, and continued until seven o'clock Friday morning. At that point, the negotiators delivered back to the Danes a revised version of the Accord. Then the leaders of this select group began arriving at eight-thirty in the morning to study it.

So Wills was a bit bleary-eyed when Chinese envoy He Yafei, a career diplomat and the vice-minister of foreign affairs, entered the room. Word soon spread that Wen Jiabao was not coming to this final attempt at negotiating a treaty. The reason? "One of the Chinese told me that they never received an invitation," Wills says, adding that he couldn't confirm it. In other words, he didn't know whether to believe the Chinese or not. It seemed to him implausible that the biggest polluter in the world was not invited to the climactic meeting of the most important climate change conference ever held. On the other hand, given the utter chaos of COP 15, not much surprised him anymore. The Danes had organized the meeting, so it was up to them to send out the invitations. "[Wen] was supposed to have been invited, but he wasn't," a senior European Union official says. "So that caused a lot of problems."

The Danes had certainly intended to invite the Chinese. After all, without them there could be no climate change agreement. They wondered aloud if the invitation had somehow got lost in the mail, so to speak, and ended up in the wrong hands within the Chinese delegation. Or perhaps there had been a mix-up inside the Danish government and the invitation hadn't been sent. Or perhaps it was sabotaged. Nobody seemed to know what had happened and there wasn't time to investigate.

In any other setting, the mystery of the errant invitation might have been regarded as a small thing, a diplomatic snafu easily smoothed over. But nothing was a small thing in these climate change talks, where distrust prevailed and the tiniest nuance merited endless discussion. Wills was more than familiar with intrigue. A 53-year-old white South African whose parents were blacklisted for their anti-apartheid views, Wills himself had been part of the underground struggle. During the last years of apartheid he worked as a conservation officer in a game reserve on the border with Mozambique and transported weapons across the border for the military wing of the African National Congress. When white minority rule ended in 1994, he became a senior administrator in various government departments, including environment and agriculture. Then, in 2002, he became South Africa's negotiator at the World Summit on Sustainable Development, with a mandate to negotiate a climate change deal, which he had been working on ever since. The freedom fighter turned professional administrator became a diplomat and negotiator. Then the intrigue began again. It seemed to have turned his long hair, which he habitually tied back in a ponytail, gray.

The meeting of the leaders, except for China's premier, continued until eleven-thirty that morning, when they handed the text back to the negotiators to have them try to iron out the still-substantial differences. Wills and the others worked on it until just after two o'clock, striving to forge a compromise. Then the world's leaders reconvened.

Outside the confines of these meetings, leaders and diplomats from other nations had little or no idea what was going on. Many countries stationed representatives in the corridors to try to pick up whatever intelligence they could.

This outside-inside scene impressed on Connie Hedegaard the great divide the Danes had created, and she realized to her dismay that she was presiding over a failure. When the conference began, she had been mildly optimistic. "We knew it was going to be extremely difficult. But I must say that in the first few days of Copenhagen I asked some of the very most experienced in this field, 'How do you feel about the atmosphere?' and the response was, 'People want to get this done.'" Many countries had announced ambitious domestic targets to show their goodwill. Ten days later, that goodwill had turned to anger, frustration and distrust. Inside the loop were the world's great powers, impatiently hammering together their agreement. Outside was the rest of the world, once again sidelined by the powerful.

More than once Hedegaard had cautioned her prime minister about thinking he could just cut a deal with an inner circle of fellow heads of government and the rest of the world would follow. She had seen what was coming in his meeting with European and American leaders, where there was a broad determination just to make this thing work, as if years of negotiations in the end meant little, as if the leaders could just "chop over" the interests of weaker countries, as Hedegaard puts it, as if their differences were not differences at all but just negotiating positions. As she sat watching the process unravel, she understood that "even in that room with all that power there, there was a limit to how much could be done."

In fact, the Danes had done what they had set out to do: they had completely short-circuited the formal UNFCCC process. The consequences would be dreadful. They had not, however, acted alone. They had plenty of help.

———

Even before Copenhagen, the talks had been stalemated. The rich countries—particularly the United States, Canada and Australia—had committed to reduction targets that so far were 75 percent below what scientists on the International Panel on Climate Change had recommended.

China and India still refused to make reduction commitments, fearing these would compromise the growth of their economies, which they still considered fragile, and their ability to eradicate poverty. Moreover, these developing countries wanted their fair share of the "atmospheric space" so long dominated by the industrialized nations. For them, it wasn't a question of the developing and emerging economies pulling back. This was their time to shine, and they wanted the space in which to do it. So for them it was strictly a matter of the rich countries continuing to take up too much atmospheric space. Essentially, both sides, rich and almost rich, refused to trade off economic growth. Neither side would budge. It wasn't so much that they were standing firm for business as usual; they just didn't know how to break a logjam that, particularly for the Americans, had its roots in domestic politics.

The leaders did not want to return home from Copenhagen empty-handed. That would only reinforce the influence of deniers and grind down even further the slow movement towards a global solution. They had to show some progress. So they needed some kind of document. Unfortunately, whatever goodwill there might have been was squandered by the flow of events and the onslaught of the clock.

Three months before Copenhagen, the Danish government had begun back-channel negotiations with world leaders on a "Danish text." This was the text that Hedegaard had warned should not be written until the second week of the conference, when the various positions would be fully clarified. But the Danish prime minister ignored his climate change minister, and months before the conference had even begun, he had worked out the framework of a secret text with the United States and Britain. Then he made the mistake of distributing it among several other countries.

Two months before Copenhagen, negotiators met for two weeks in Bangkok. Rumors of the text spread. The result was total stalemate. Developing countries at one point threatened to walk out.

One month before Copenhagen, at a meeting in Barcelona, organizers hoped to smooth out the bad feelings and come up with the essential bones of an agreement on both the Kyoto Protocol amendments and the agreement on Long-term Cooperative Action, but that meeting also ended in deadlock.

The Danish plan to raise awareness of this "decisive" conference to the point where every world leader would want to attend had worked, and the Danes hoped that the presence of so many leaders in Copenhagen would at last break logjams and force compromises that would lead to a final global accord to fight climate change.

That was the hope. What resulted more nearly resembled a circus.

Trust is a key ingredient in any United Nations agreement on climate change. Each country has one vote and the ultimate quest is for a consensus. The Danes' diplomacy, however, had been heavy-handed, and as crunch time approached, rumors persisted that the Danes were trying to impose a "Danish text" on the convention.

Two days into the convention, the *Guardian* ran a story on what it claimed was the secret Danish text, of which the paper had obtained a leaked copy. At that point, however, the secrecy of it all had long since vanished. The Danes had distributed copies at closed-door meetings several weeks prior to Copenhagen. Warnings that none of the copies should leave the room went unheeded. The text soon found wider distribution, which is how a copy ended up in the hands of the British newspaper. The developing countries immediately cried foul, claiming that the text favored the industrialized countries and was an attempt by them to impose one agreement on the conference and thus destroy the Kyoto Protocol. In fact, the text published in the *Guardian* was incomplete, and did not include the Danish proposal for a renewed Kyoto Protocol.

After the *Guardian* story ran, the conference dissolved into bickering and recriminations. Countries such as India and China, who had expressed no willingness to reduce their emissions in the short term, were joined by African and South American countries in denouncing the text as an attempt by the rich countries to usurp the United Nations process and impose a last-minute agreement that had been negotiated in secret. The Danes denied there was a secret text. They were right: the text had been shown to all the major negotiators well in advance of the convention. What's more, Reuters had run a story about it eight days before the *Guardian*'s "leak." Still, the frenzied protests over the "Danish text" and the loss of trust—primarily from China, India and Sudan—plunged the convention into disarray. Indian environment minister Jairam Ramesh, three months later, said the "leak . . . destroyed Copenhagen from day one."

There is, however, little evidence that this is true. The so-called leak of a "secret" document of which most parties were already aware was no doubt purposely used by countries such as India and China to heighten the atmosphere of confrontation, confusion and distrust, which was in turn relayed to home audiences by the five thousand journalists who covered the conference. The tense situation allowed these countries to escape blame for destroying the climate change agenda. The fact is, the diplomats in Copenhagen were too sophisticated to allow a simple leak to jeopardize the talks unless they wanted it to. Still, the ultimate failure to reach any agreement at Copenhagen was a result of both poor Danish diplomacy and Danish disorganization.

The schedule was extremely tight. By the end of the week of talks, negotiators were supposed to have produced texts that were mature enough for the environment ministers—due to arrive at the end of the week—to begin studying. Those ministers then had a mere five days to negotiate the final terms of the agreements before the leaders arrived towards the end of the conference to put their signatures on a done deal. But with the ministers and then the heads of state and government due to arrive, everyone kept passing the buck. Alf Wills recalls, "The negotiators didn't have the mandate to change their positions. The ministers didn't feel like they had the authority to change positions because the heads of state were coming. And then there were only two days left and there was no time for the heads of state to swing the deal in an inclusive way. And so the Danes were forced into a small-group approach that was not acceptable."

On the Wednesday of the final week, as the 120 world leaders began landing in Copenhagen, Danish prime minister Rasmussen took over from Connie Hedegaard as president of the conference. Since the negotiations would now be in the hands of prime ministers and presidents, it was protocol that the head of Denmark's government should lead the way. In her final act as president, Hedegaard announced the changeover, paused for a moment, then told the plenary that Rasmussen would be presenting a new Danish text to the convention. Only her hesitant voice revealed the backstage struggle that had gone on over how and when to present this text. She had thought that releasing the text was still premature because the talks had not advanced far enough. "It was only a question of if you do it prematurely, then you risk getting stomped," she says, but Rasmussen insisted it was the only way to get the process moving. As soon as she mentioned the text, she noticed many delegates reach for their microphones, hesitate and then sit back. She realized they were waiting for Rasmussen to take the president's chair before they pounced.

As soon as Rasmussen, round-faced and boyish-looking at forty-six, slipped into his president's chair at the head of the conference, the Chinese delegation attacked: "This is a very grave issue and this has something to do with the trust between parties and also the trust between the host country and the parties . . . This way of proceeding is not respecting the parties and we are opposed to this proceeding."

Sudan, speaking on behalf of the G77, agreed. "It is not respectful to us to be expected to rubber stamp a text coming from out of the blue." The Brazilians and South Africans joined in the condemnation, as did Bolivia and Venezuela.

Rasmussen countered, "The world is expecting us to reach some kind of agreement concerning climate change and not just to continue discussing procedure, procedure, procedure."

The Chinese didn't back down: "I think the matter is not just procedural, procedural, procedural. There are very serious issues of substance. The respect of the host for the 192 parties. This is a party-driven process. You can't just put forward some text from the sky."

Rasmussen: "It is not our intent to put any text on the table from the sky . . . Whatever we must decide to put on the table in order to get things moving . . . would be papers that fully respect the outcome of the negotiations among the parties."

China: "Unfortunately the presidency put forward something from the sky, I should say, a parachuted text . . . We should stick to our mandate and not work on some hidden agenda and try to impose some external thing upon the distinguished parties."

Rasmussen appeared totally flummoxed even though Hedegaard, the staff of Denmark's climate ministry, as well as Yvo de Boer, the executive secretary of the UNFCCC, had all warned him about the dangerous optics of trying to present a text from the conference presidency to a United Nations forum. At last he realized his mistake, and quickly sounded the retreat. "At this moment we haven't presented a text from the presidency . . . I really urge you to accept that we now start the list of speakers. I hope that the future presence of so many heads of states and governments in Copenhagen will pave the way for a successful outcome."

The damage, however, was irreversible. His attempt to mollify the parties failed and the battering of the Danish presidency continued for another thirty-three minutes. Then any hope of a revival was stymied by the leaders themselves, who insisted (once they arrived) on delivering official speeches to the whole convention.

Part of the problem was that the Danes simply hadn't done the math. That evening—with only three days remaining—the various leaders began their speeches to the plenary. Each was given a ten-minute slot. Some spoke less, some more. At the outside, there were twenty hours of speeches scheduled, plus the time it took for each speaker to leave the podium and the next one to take his place. Most of the countries trailed their own media representatives, who took extra minutes setting up their cameras so they could broadcast the speech back home. Venezuelan president Hugo Chávez, sporting his trademark dark suit and bright red tie, spoke for twenty-six minutes on the evils of capitalism and America, and that was after he took an extra ten minutes just to get to the podium because he was mobbed by fans, many of whom wanted his autograph or their picture taken with him or both. After the speeches, there would be only about forty-eight hours left in the negotiations, much less if delegates decided to catch some sleep and a bite to eat. The United Nations rules stated that the plenary sessions had to be held one at a time. So while the leaders spoke—often to a sea of empty chairs—the supreme body of the convention could not meet to discuss the two treaty texts. The Danes could have obtained a special mandate from the plenary to continue the negotiations while the leaders spoke, but they never did. The reason was simple: they had already begun secret negotiations around a small group of world leaders led by the

Americans and Chinese, called the Copenhagen Commitment Circle. In a word, the Danes had sabotaged their own meeting.

Zammit Cutajar recalls that by the second week it seemed as if the Danes were only going through the motions of holding a conference. Connie Hedegaard had set up various contact groups outside the plenary to continue negotiations on several issues while the leaders spoke. "I half rashly agreed to chair one of the groups on shared vision," Cutajar says. "It was clear that some of the people weren't serious. The Americans, for example, were putting down an amendment and you could tell from the way the guy was putting it down that his only intention was not to reach an agreement, because his boss was negotiating in another room and there was no point having an agreement at his level of official. So that was a complete waste of time."

At this point, Cutajar decided to produce a revised Long-term Cooperative Action text himself as a possible consensus document that could be presented to the ministers. Many welcomed it as an attempt to get the negotiations back on track. But not the United States. Their delegation blew up because the text proposed that each country commit to emission reduction targets. The Americans wanted no such treaty constraints.

The contact groups that Hedegaard set up had their first meetings, shared out the work and then never met again. They were left hanging in the air. Hedegaard also set up working groups where environment ministers were to meet on Monday and Tuesday of the second week to examine key issues such as setting emission targets and financing projects to help poor countries adapt to climate change. "The ministers did damn all with them," Cutajar says. "Over those two days those ministerial consultations produced absolutely nothing. Absolutely nothing . . . It was a very messy situation. For me it was a cover for the fact that there was something shady going on . . . It was a cover for the fact that some things were starting to happen behind the scenes."

By mid-morning on the last Thursday, delegates were beginning to panic. Time was running out and there was no sign the
leaders were ready to stop speaking to the plenary session. The
amendments to the Kyoto Protocol alone ran forty-two pages,
with at least 120 points still undecided. The text for a treaty on
Long-term Cooperative Action, which would commit all countries to post-2020 actions as well as include the United States in
short-term actions, had been reduced to only ten pages, but there
were 67 points still undecided. More than 50 percent of the text
was bracketed in options. In addition, there were separate submis-
sions for amendments on the table from the African countries as
well as Bolivia and Peru and the 43-member Association of Small
Island States. Furthermore, the industrial nations were still pushing for one treaty to merge the Kyoto Protocol and the convention on Long-term Cooperative Action into one global agreement.
Moreover, nobody trusted that the United States would ever ratify
a treaty no matter what form it took.

Desperate to get the talks back on track, the G77 invited de
Boer on Thursday morning to explain what he could do to persuade the leaders to stop speechifying so they could get on with
the negotiations. Unfortunately, he said, his hands were tied; the
best he could do was to try to persuade the leaders to take a
breather at about eleven o'clock that morning to allow the negotiators to meet in plenary session. But not even he could stop the
leaders from talking. One hundred and twenty world leaders had
shown up at the biggest global conference in history to attempt
to reach an agreement on what they all agreed is "one of the greatest challenges of our time" and they refused to curb their rhetoric
long enough to negotiate a deal.

At the state banquet held on Thursday evening, the Danish
prime minister requested a special meeting of a select group of about
forty leaders on the final Friday. Included would be all the major
polluters and regional representatives. The meeting would be held

in Rasmussen's offices at the conference center. The topic would be a new draft text with thirty-two clauses—a proposed "Copenhagen Accord." The scheduled time to meet was eight o'clock the next morning. Formal invitations were sent to the leaders.

That's when things went from bad to worse. The Chinese invitation never arrived. The problems that had hamstrung the conference as a whole went viral and infected the back rooms.

The Chinese felt they had been snubbed, while the leaders of the industrialized nations, who didn't realize that the Danes had failed to convey an invitation to Chinese premier Wen Jiabao, thought the Chinese were snubbing them. For the Chinese, the errant invitation served as proof of Western duplicity. The Chinese bywords were "transparency and democracy," not concepts Chinese Communist Party members are often clear on. But they were useful weapons to galvanize environmental groups and the poor and developing countries and bring them to China's side. It may be that the Danes "lost" Wen's invitation because the Chinese had refused to participate in an earlier "friends of the Chair" meeting. Their refusal angered Prime Minister Rasmussen, who saw it as a sign that the Chinese didn't want a deal in Copenhagen. Whatever happened, the Chinese confounded the problem by sending two low-level envoys to the Friday morning meeting. The industrialized nations, unaware of the invitation snafu, regarded it as a direct snub.

Inside the meeting, world leaders found themselves doing what their negotiators were supposed to have already accomplished, debating every word, every line, every punctuation mark. Tensions rose and the process became painfully laborious as one of the two Chinese envoys, He Yafei, constantly stopped the talks to consult with his boss, Wen Jiabao, who was holed up in another room— further annoying the Western leaders, who wondered why Wen wasn't there. "It would be nice to negotiate with somebody who can make political decisions," President Obama commented. Yet, Wills recalls, though Sarkozy and Angela Merkel were annoyed, Obama didn't seem upset that he had to negotiate with Chinese underlings. "It was clear he just wanted to get the thing done," he says. Obama strove to bring the parties to agreement. "He tried real hard," Wills insists. "You saw him in the corners . . . trying to make deals . . . There were certain . . . difficult areas, which had to be solved bilaterally with parties that were interested in those particular issues. President Obama was definitely involved in some of the key ones, like the forest issue, like the issue . . . around the nature of the commitment of developed countries, how to measure action from developing countries, how to verify it."

The Chinese rejected some of the most fundamental elements of the proposed agreement. For instance, they didn't want any reference to long-term emission targets, even by the rich countries, which flabbergasted Chancellor Merkel. The original draft recommended 50 percent reductions by 2050. The Chinese, however, weren't satisfied with simply putting down a figure; they wanted an assurance that whatever reduction targets were to be imposed, in the end there would be an equal share of the carbon space for all countries and that the share would be based on population. Major industrialized countries such as the United States, Canada and Australia, which emit between 17 and 23 tons per capita, would have to yield large swaths of atmospheric space to the enormous populations of China and India, which emit only 5.5 and 1.5 tons per capita respectively.

"The Chinese were pushing for the solution of the big conundrum, which was the sharing of the atmospheric space, and the developed countries refused to engage," Wills says. China basically wanted to hold the rich countries' collective feet to the fire on short- and medium-term goals, and Wen never budged from this position. For their part, the industrialized countries would not agree to ambitious short-term targets unless China and developing countries agreed to long-term targets.

The problem for the developing countries is not simply future emissions; it's the cumulative emissions already in the atmosphere that have been put there by industrialized countries. Despite their rising emissions, China and other emerging nations still consider themselves small players relative to the U.S.A. in historical terms. "In the context of a differentiated responsibility for the past, which the Americans refuse to acknowledge, the industrialized countries have to pull back and give space to the developing countries," Wills says. "That's the fundamental issue. And that is the new world order issue. And that has not changed. Not yet."

After two hours of word-by-word, line-by-line negotiations, during which Angela Merkel constantly but unsuccessfully pressed both China and the United States to show more ambition, one major point remained. The Americans and the Europeans insisted that China—and all developing countries—agree to some form of international monitoring, reporting and verification (MRV) of their emission reductions. The Chinese had already committed to various degrees of monitoring two months earlier, at a bilateral meeting with the Americans in Beijing. In Copenhagen, however, Obama pushed for stronger reporting commitments from the BASIC countries (China, India, Brazil and South Africa), which were the main emerging economies. He had said publicly that lack of a Chinese commitment to international MRV was a deal-breaker. It was mere pantomime. Obama consulted again with British prime minister Gordon Brown, French president Nicolas Sarkozy and Japanese prime minister Yukio Hatoyama. They agreed that they had to tackle each of the BASIC countries separately on the issue, especially China.

Obama went down the hall looking for Wen Jiabao. As he approached the room, two Chinese officials asked him to wait, because Wen Jiabao was not ready for him. Obama and his team pushed by them and went into the room. Waiting for them was a united front of leaders of emerging economies: South African president Jacob Zuma, Brazilian president Luiz Inácio Lula da Silva, Indian prime minister Manmohan Singh and, of course, Wen Jiabao. If there had been any attempt to divide and conquer, the Chinese had outmaneuvered the Americans.

China's meeting room was now packed with advisers. Obama had brought a team of about ten people, including his chief negotiator, Todd Stern, and Michael Froman, his deputy national security adviser for international economic affairs, as well as Secretary of State Hillary Clinton. Wills, who was sitting with his minister, watched the Americans' every move as they faced off against the BASIC nations.

With all talk of targets off the table, Obama focused on international monitoring, reporting and verification of every country's emission reduction efforts. The Americans would later make it sound as though they had scored a major victory in getting an agreement from China, but this was not the case; China didn't sign on to anything to which it had not already committed itself. Just a day earlier, Wen had reiterated China's position on MRV to the media. He committed China to a domestic MRV program, one that is legally binding and transparent, and added that China "will consider international exchange, dialogue and cooperation that is not intrusive and does not infringe upon China's sovereignty." He said China would submit to international MRV on a project receiving international funds. This is something China already accepts when it sells emission reduction credits on the European carbon market to finance hundreds of hydroelectric, geothermal, solar and wind energy projects and other so-called emission reduction projects. From 2007 to the end of 2009, China used 172.9 million credits[2] or about one-third of the world's issued emission credits, making it one of the largest players in the European carbon market, which has been shaken by allegations of widespread fraud. The vast majority were bought by Canada, Japan and European countries, meaning they were spending money to reduce emissions in China while they increased their own emissions.

Wills says that none of the other leaders in the room trusted the Americans. "The problem is that the U.S. will not sign up to anything unless they have their legislation through the Senate. So the question is when is that going to happen. We don't want the experience of Kyoto, where the U.S. negotiated Kyoto and then subsequently didn't ratify."

The Obama administration was well aware of this distrust, but could do little but offer vague assurances that the Senate would come round. Todd Stern had been negotiating with China on and off throughout 2009, though he was rarely present at any of the official UNFCCC negotiating sessions in Bonn, Bangkok and Barcelona. He had left that to the precise and unbending Jonathan Pershing, preferring backroom negotiations with the world's major polluters to the murkier, often chaotic UN negotiations.

If Stern were a cowboy, he would be nicknamed Slim. Tall and thin, with pinched cheeks and narrow eyes, rimless glasses and short-cropped gray hair, Stern is a Chicago native with a history degree from Dartmouth and a law degree from Harvard. He worked on former U.S. president Bill Clinton's election campaigns, then served as a White House staff secretary and chief climate change negotiator in Clinton's administration. When George W. Bush won the presidency and rejected American participation in a global climate deal, Stern disappeared into the fringes of Democratic Washington, working as a partner in the public policy section of a corporate law firm and as a senior fellow at the Center for American Progress, a liberal think tank.

In 2007, with the Democrats looking as if they would take back the White House, Stern and a colleague named William Antholis at the Brookings Institute wrote a twelve-page memo calling on all presidential candidates to support urgent action to combat climate change by stabilizing greenhouse gas emissions. What was notable was Stern's opposition to the United Nations approach to negotiating a climate change treaty. He claimed the

UNFCCC was hamstrung by structural deficiencies that made a global agreement impossible. He wanted negotiations to center around a sort of polluters' elite:

> The negotiations typically involve more than 150 nations, often grouped into competing blocs, which must negotiate internally as well as among themselves. Environmental issues rarely motivate top-level decision-makers, so negotiations are conducted by bureaucrats empowered to make technical level decisions, but not political compromises among countries. Moreover, substantial UN bureaucracies often grow up around the issues, further slowing the pace of progress. When political energy is finally brought to bear by ministers, arriving with too little time left in a session, the results are too often continued stalemate or modest compromise. And that doesn't count the difficulties of getting a signed treaty ratified, or ensuring the implementation of a treaty once it has entered into force. This is no way to run a planet.

The way to run a planet, according to Stern, is to concentrate negotiations in the hands of an international green team, an ecological board of directors selected both for their "political standing and for the importance of their ecological roles." Stern suggested that three developed and four developing countries plus the EU make up the "E8": the United States, the European Union, Japan, Russia, Brazil, South Africa, India and China. He considered adding other players—such as Canada (world polluter number seven), Australia, Indonesia or Mexico—but decided against it because "preserving a sense of intimacy and informality argues against a larger grouping." The E8 would represent such a large ecological footprint—74 percent of global man-made GHG emissions—that its actions "would be consequential in their own right and could set the terms of the policy debate more broadly, whether within existing environmental conventions or outside of them."

Stern said the E8's two main goals would be to tackle climate change and deforestation. Included in its mandate would be setting national emission limits and negotiating deals on such things as technology development and transfers. In forestry, where the E8 would represent half the world's forests and the largest wood consumers, the group would take steps to "shut down the robust trade in illegal logs; support a strong certification program and consumer boycott of products made from uncertified wood; eliminate perverse subsidies that accelerate forest loss; and plan how the vast appetite for wood in the E8 countries should be managed." Once the E8 had decided what worked for its members, it would then try to persuade other nations to sign on to its agreements. Stern was confident the E8 would eclipse the "torpid pace" of the United Nations negotiations, where consensus had to be found among the vested interests of 192 countries. His goal was to hold the first E8 meeting on April 22, 2010, which, he pointed out, was the fortieth anniversary of the first Earth Day.

Stern's vision, of course, was unworkable, not least because the other 160 countries might object to the dictates of the new world order represented by the E8, no matter how well intentioned. Yet it reflected Stern's and the United States' frustration at being unable to control a process that colored every aspect of a nation's fabric and presented nothing less than a revolution in political, economic and social thinking. It's probably not surprising that a superpower which for sixty years had ruled the world's economy would seek to take refuge from the winds of change in a small group through which it could "set the terms of the policy debate." That Stern, a seemingly intelligent man, could propose such a naive solution to the "torpid pace" of the United Nations negotiations ultimately reflected a failure on the part of the United

States to grasp the fact that the world was moving ahead of the Americans. While America's about-face on Kyoto was still embedded in the collective memory of other nations, for Stern it was ancient history. When confronted by America's past performance, his oft-repeated response was: "The past is the past. We have to move on." But it is the Americans who don't understand what moving on means.

After Obama was elected and had appointed Stern as his chief negotiator, there were clear signs that his new administration realized it could not reverse American intransigence on climate change. Senate opposition was still too powerful. Even with the Democrats in control, there was no realistic hope that Obama could get the needed two-thirds Senate vote to approve an international treaty. Stern was attached to the State Department, which oversees all foreign negotiations and has functional bureaus that handle such topics as climate change. The climate change bureau is a vortex that sucks in information from all other departments and is the center of a storm of competing interests. It weighs them, teases out every imaginable eventuality and hammers out strategy. Stern handles the political side and, while he reports to Hillary Clinton, he has a good deal of independence in steering his team.

The Americans at Copenhagen were well aware that distrust in the United States' ability to sign up to a legally binding agreement was the central roadblock to an international treaty. Without trust in America, China would not play the game. Nor would Japan or Canada. The Americans' solution was to ignore the trust issue, finesse their way around historical responsibility and try to gain commitments to something less formal than an international treaty.

When Stern came to his first UNFCCC meeting in Bonn in March 2009, and received a standing ovation from the nations of the world for ending the eight-year deniers' drought of the Bush administration, his real goal was to establish close ties with his E8 and use them to help fashion a more informal international agreement that would not require Senate treaty approval. The United States would show its commitment not by its signature but by its actions.

By Copenhagen, the ovation given the American was a distant memory. Stern's strategy had failed. No country could ignore the United States' climate history and its domestic political reality. Stern's main target, China, simply did not believe the Americans could live up to any of their commitments. China also believed that the U.S.A. had a historical responsibility to reduce its carbon emissions more than any other country as well as surrender the atmospheric space to countries such as China. The Americans, of course, flatly rejected both demands. In addition, China's national interests aligned with its fellow emerging economies and with the poor and developing countries, such as those in Africa, whose considerable natural resources China coveted and who all favored a strong international treaty. China, like most countries, still believed that an international treaty was critical to meeting the threat of climate change. They were all aware of the necessity to cut emissions on a global scale. Thus, the efforts to obtain some kind of deal in Copenhagen were not all show. The problem was that trust in America, which held the key to a global treaty, was just about zero.

By gate-crashing the meeting between China and the other BASIC countries, Obama confronted four of Stern's imagined E8. Together with the United States, they represented 50 percent of the world's greenhouse gas emissions. Add the other three potential members—Europe, Russia and Japan, who were waiting in the other back room—and you had almost 77 percent (slightly higher than when Stern wrote his E8 strategy). Obama had come into the room to work out a clause on monitoring, reporting and verification—his "deal-breaker" clause. After much discussion, he agreed to what China had promised a month earlier at their meeting in Beijing: China would accept MRV on projects supported by international funding; otherwise China would police itself and the world could see its reports.

With that commitment in hand, Obama emerged from the meeting waving the Copenhagen Accord before a hand-picked group of reporters, claiming that he had orchestrated what he called a "meaningful and unprecedented breakthrough." He then dashed off to Air Force One and flew home. The confirmation was Obama's and his alone. The contents of the Accord, as well as the 160 countries left out of the process, would ultimately say otherwise.

The Copenhagen Accord had shrunk to a mere twelve paragraphs, from thirty-two. It proposed keeping mean temperature rises below 2 degrees Celsius without actually making any commitment. There were no references to a global emission reduction target. The Accord proposed that both developed and developing countries submit targets, but it didn't specify what those targets should be. It proposed the creation of a $30-billion fast-track financing fund to the end of 2012 to help poor and vulnerable countries adapt to climate change, as well as a $100-billion annual international fund for financing post-2020 mitigation and adaptation, without saying where the money would come from or how it would be administered. The Accord also called for a program to reduce deforestation, which the Intergovernmental Panel on Climate Change (IPCC) has concluded is responsible for up to 20 percent of the increased carbon in the atmosphere. China pushed for and got a reassessment period built into the Accord requiring parties to review by 2015 whether the goal should be to keep the global temperature rise below 1.5 degrees Celsius. As it stands, however, scientists say we have already emitted enough greenhouse gas into the atmosphere to increase the mean temperature above the 2 degrees Celsius limit, never mind 1.5, probably within the next fifty years.

The Accord reflected the least ambitious negotiating positions of the most flagrant polluters. For countries such as Canada and Saudi Arabia, for example, there would be no rush towards alternate energy sources. For them and other major oil producers, these negotiations were about long-term security of demand for their oil. For countries such as the United States and China, the Accord meant business as usual. For the poor countries and the island states, it meant that adapting to climate change would just get harder. "To the Obamas of this world, this seems particularly okay," Zammit Cutajar observes.

Jonathan Pershing later described the Copenhagen Accord as just a step along the way to something better. "My sense is we need to be careful about taking all of our steps at once," he told a meeting in Washington soon after Copenhagen. "You can't eat the apple in one bite. You take a bunch of bites of the apple."

The step-by-step process might make sense if you were talking about reducing obstacles to trade, for example, but it makes little sense in the context of climate science. Melting glaciers, ice sheets and ice caps as well as rising sea levels, dying coral reefs and altering weather patterns do not respond to the art of the politically possible. Climate change is not waiting for us to digest the apple one bite at a time.

In the early hours of Saturday morning, the delegates from 192 countries gathered to decide on the Accord's fate. Most world leaders had already flown home, leaving their ministers to write the final act of Copenhagen. As delegates entered the hall, there was no sense of unity, no feeling of consensus. If there was a spirit, it was an angry one.

The United Nations climate change negotiations do not seek unanimity; they seek a consensus that reflects a general absence of opposition. When the parties adopted the Kyoto Protocol in 1997, the Saudi Arabians and their OPEC supporters waved their flags in opposition. But the president of that convention ignored them and gaveled the Protocol through. The vast majority of the world's nations stood up and their wild applause drowned out anything the Saudis could muster.

The debate in Copenhagen was confrontational and continued well into Saturday morning. The Danish president lacked the credibility needed to bring the delegates together. Eventually, most countries appeared ready to hold their noses. "There was nothing binding in approving it," one delegate told me. "It was just a political agreement showing good intentions, I guess." In other words, there wasn't much apple in the apple. The Accord would be largely meaningless, so why not vote for it.

But any hope that Prime Minister Rasmussen could gain a consensus was lost when he repeatedly announced to the convention that there was no consensus. "I don't see a consensus," he kept saying. "We don't have consensus." He looked around the hall, shrugging.

Cutajar sat on a folding metal chair, his coat draped over his lap, among deserted exhibition booths in one of the Bella Center's many convention halls, watching Rasmussen on a television monitor. It was five in the morning when I joined him. He shook his head in disbelief. "He is confusing the lack of consensus with the seeking of consensus, which is what the process is all about," he said. "There is no hope for a consensus now." Zammit Cutajar looked spent.

The debate over what to do with the Copenhagen Accord was testament to a creeping conformity. The attitude was, "Look, just go along with this, okay. We can't walk away empty-handed. We'll fix it later."

That was generally the tone of the discussion as once again the rich faced off against the poorer countries, which have spent the last four centuries conforming to a European vision of the world. The developing world, however, appeared to be drawing a line. Or kind of. When money is promised, it is never that simple.

The forty-one industrialized countries wanted the convention to approve the Copenhagen Accord as the basis for a treaty to be signed a year later in Cancún. The developing countries found the Accord entirely too weak. Yet at the same time they did not want to leave Copenhagen with nothing. If that happened, delegates feared the process would be severely damaged in the eyes of the public and faith in it would vanish.

While the BASIC countries supported the Accord, they largely sat on the sidelines during the final debate. The rest of the G77, however, was fairly solid in its opposition.

The election of Sudan to the chair of the G77 is typical of the many levels of intrigue that are brought to the climate change table. Sudan won the vote on September 26, 2008, with the support of China, and held the position throughout the 2009 negotiations. For China, Sudan's election was purposeful. Sudan's major export is oil and its best customer is China, which now controls most of the major oil wells in the country—purchased from Canadian companies—as well as the pipelines, refineries, port facilities and tankers that ship the crude home. Swedish, French, Italian and British petroleum companies also have interests in Sudan's considerable oil reserves, but compared to China they are small. China, Russia, Poland and Bulgaria have sold light, medium and heavy weapons to Sudan—with China being the current supplier of choice—which the Muslim-led government has used against the southern black Christian populations and the people of Darfur in the western part of the country. Part of the conflict in the south is over control of the oil region and the creation of a sort of "sanitized" area for oil workers. The turmoil also helps to keep the price of oil high, which is beneficial to OPEC and other oil producers but not to most members of the G77, whose treasuries are emptied by high-priced oil. China, on the other hand, often pays for its resources in Africa by building infrastructure with cheap Chinese labor. Sudan's president, Omar al-Bashir, is under indictment by the International Criminal Court for crimes against humanity and war crimes in Darfur, where he is accused of arming Arab nomadic tribesmen called the Janjaweed to drive out the black African farmers. Estimates of genocidal killings range widely from 20,000 to 400,000. The United States has listed Sudan since 1993 as a state sponsor of terrorists.

Lumumba Stanislaus-Kaw Di-Aping, as Sudan's UN ambassador and chief climate change negotiator, was the voice of the G77 at Copenhagen. He became one of the more outspoken and passionate players in the negotiations and made few friends along the way, particularly when he reminded the West of its colonialist, slave-trading history. A Christian, he comes from oil-rich southern Sudan, which has for decades been at war with the Islamic government in Khartoum. His mother was Dinka, a people who were enslaved by the Arabs, and his father was a Lwo tribal chieftain, politician and general. The Stanislaus in his name comes from Polish missionaries. He is of medium height and slender, dresses like an Englishman and lives in New York. Sweden's chief negotiator, Anders Turesson, who often speaks on behalf of the EU, liked to refer to him as the "New Yorker from Sudan." Lumumba did his undergraduate degree in Sudan and went to Oxford on a Commonwealth scholarship, where he studied economics. He then worked for the management consultant firm of McKinsey and Company in London and South Africa before becoming a diplomat. An economist by training, he says his main interest in these negotiations is finance. He is a hard man to pin down. In a brief conversation he once told me, "Oil has shaped a lot of my politics, but you will have to go through a long discussion in order to decipher your own reading of what I make out of that." He promised on several occasions to give me an interview, never showed up and then apologized profusely when I nabbed him in the hallways. When he speaks, it's with a soft, slow precision that gives the listener the impression he is choosing his words carefully, that he's saying something terribly important, but you never know. Sometimes his words get lost in the waiting.

"Let me be very blunt," he said at a press conference in Copenhagen where he was reacting to the offer from Europe and the United States of $10 billion for fast-track financing to help poor countries adapt to climate change. "Ten billion dollars would not buy developing countries' people enough coffins. It is as simple as that."

Journalist #1: What? It's a little hard to hear. Could you repeat it.
Lumumba: I said $10 billion would not buy developing countries'
    citizens enough coffins.
Journalist #2 (amid much confused rumbling): Coffee?
Journalist #3: Why do they want coffee?
Journalist #1: No. Coffins.
Journalist #4: Coffins?
Lumumba: It's the thing you bury people in.

Lumumba's partner at the negotiating table was the equally persistent and uncompromising veteran Filipino diplomat Bernarditas de Castro-Muller. She has spent much of her adult life in Geneva, where she lives with her Swiss economist husband. The daughter of a general who was head of the Philippines' police commission, Castro-Muller is known by her adversaries as the "dragon lady" and by her friends simply as Dita. Her tiny stature and broad smile mask a fierce determination to fight for the rights of developing nations. A career diplomat, she has represented the Philippines since 1986 in multilateral negotiations on various environmental issues, such as the decline in species and desertification, and then became involved in negotiations for the 1992 UNFCCC. Like many of the negotiators, she was trained in the science of climate change by the United Kingdom's Hadley Centre, and since the early 1990s she has been working on climate change negotiations full-time. So she has a long institutional memory and never hesitates to remind her fellow delegates of something they said or promised years earlier. She has always insisted on holding the rich countries accountable, demanding they meet their Kyoto commitments, reduce their emissions by up to 40 percent by 2020 and pay upwards of $400 billion a year to help poor and developing countries adjust. Her job within the G77 is to help craft common positions among the members.

We met in a restaurant in Bonn for an interview. "Developing countries individually don't count for much," Castro-Muller said. "If I was going to pursue my position on my own, I would have to be a very, very, very strong voice. The strength of the group is in its unity. The diversity allows us to have a bigger view of the issues. In many, many ways, individually, many of us are taken advantage of. The developed countries are much more coordinated than we are. Why? Because they do have the possibility to meet among themselves. They have the financing capability to travel here and there among themselves. There is the OECD [Organisation for Economic Co-operation and Development], there is the European Commission. They have forums for everything. We don't have that. We have a tiny office in the United Nations, the main office of the G77, and then chapters everywhere. We are not funded. We get contributions for specific things, like we have to publish something or we have a reception or maybe a coordinated meeting. But the developing countries in general don't have a specific forum where they meet."

So I asked her if that's why she feels compelled to raise her voice above everyone else's. "In the end, what I am doing is really clarifying to everybody what their respective obligations are and how they should do this. We all agreed to do this. We ratified this convention. In the end, it all boils down to that . . . The developing and poor countries suffer the most from climate change. One typhoon and Manila is destroyed. How can you develop? How can you ask us to take targets for mitigation? What do you mitigate? And this is the case for many developing countries. That's why they say they need to develop. Even China, they have large pockets of poverty in China. And India. India is not at all in the same position as China. It is a least developed country. They have more poor people than all of Africa put together. It's just that they are big."

Her distinctive gravelly voice with its exasperated, scratchy tones reproaches the industrialized countries for their failure to meet their climate change obligations. Industrialized countries criticize her for being obstructive. Some Western delegates try to imply that she is senile or slightly unhinged. Two months before Copenhagen, her own country canceled her accreditation. The rumor was that the United States and the World Bank had applied pressure on the Philippines to have her removed from its delegation. "Some people said the U.S. did it, but I really do not know. It might be. Listen, like I said, we are negotiating our share of environmental space. That means tons of money. If people think I stand in the way, they might do it." And then she recalls a story told to her by a Caribbean colleague at the UN who had incurred the anger of the United States. "So [the Americans] went to the minister of foreign affairs and said, 'Your representative in New York is making 90 percent of his representations against the United States.' Complaining about him. And the story goes that the minister looked at the Americans and said very kindly, 'Is that so? I am really very concerned to hear that the United States is taking 90 percent of its positions against my country.'" She laughed and then took another mouthful of salad.

After her own country rejected her, the Sudanese gave Castro-Muller accreditation in their delegation so she and Lumumba could continue their work as the tag-team leaders of the G77. Her mandate now is to try to negotiate billions of dollars out of the industrialized countries to help poor and developing countries build their economies.

Her office is in Geneva, where she is an adviser to the South Center, a small think tank financed by developing countries. Its mandate is to help mobilize poor and developing countries to increase their bargaining power on the world stage. It is led by Martin Khor, a Cambridge University–trained economist from Malaysia. He believes in an uncompromising stance against the industrialized world's refusal to cut significantly its greenhouse gas emissions and to pay substantial reparations to the underdeveloped countries for polluting the atmosphere in the first place. To the West, he is another hard-liner on climate change, part of a small Geneva-based group of developing-country diplomats that also includes Angélica Navarro Llanos, who is Bolivia's lead negotiator and its United Nations ambassador to Geneva. Throughout 2009, Khor and his Geneva group tried to rally the G77 to an uncompromising position of either a good deal or no deal. The trouble was that no deal at all could bring hardship to those G77 members already feeling the effects of climate change. Fast-track financing would not flow their way and action on climate change would be delayed. At least that was the threat. So the question as the final plenary debate opened was whether the G77 would stand firm against the Copenhagen Accord.

Lumumba played the only real hand most G77 countries had: moral righteousness. He did it with uncompromising candor, but he was up against the utilitarian thinking of Western leaders. During a private meeting of African delegates midway through the conference, Adam Welz, a climate activist from South Africa, captured on Twitter fragments of Lumumba's words of encouragement to the African members of the G77:

Lumumba Di-Aping:
2 deg C is "certain death" for Africa, is "climate fascism," we are asked to sign it for $10 billion + asked to celebrate
7:41 AM Dec 8th, 2009 from web

Lumumba Di-Aping:
Think carefully what you do in the name of Africa
7:41 AM Dec 8th, 2009 from web

Lumumba Di-Aping:
Your campaign and your resolve must be unbending
7:42 AM Dec 8th, 2009 from web

Lumumba Di-Aping:
It is unfortunate that after 500 yrs + interaction with the West we are still considered disposables
7:59 AM Dec 8th, 2009 from web

Lumumba Di-Aping:
"One Africa, One Degree Celsius. Two degrees is certain death" Spread it around "why are we [Africans] being sacrificed?"
8:25 AM Dec 8th, 2009 from web

Lumumba Di-Aping:
I would rather die with my dignity than sign a deal that will channel my people into a furnace
8:28 AM Dec 8th, 2009 from web

Hillary Clinton brought to Copenhagen a promise that the United States would work with other industrialized countries to create a fund for US$100 billion a year in financing to developing countries for adaptation to climate change. This would be in addition to the $10 billion a year in short-term financing for 2010 through 2012. All paid for by industrialized countries. She emphasized that no money would go to China, even though that country had not asked for any. She also promised that America would cut its emissions to 17 percent below 2005 levels by 2020, 30 percent by 2025, 42 percent by 2030 and more than 80 percent by 2050.

"If we cannot secure the kind of strong operational accord I describe today, we know what the consequences will be for the farmer in Bangladesh, for the herder in Africa, for the family being battered by hurricanes in Central America," she said. In other words, sign or die. This is one tough-talking lady.

The European Union had earlier proposed a slightly higher amount: US$143 billion a year. But even that came nowhere near to the amount the United Nations and the World Bank claimed will be necessary to help poor countries reduce their emissions, convert their economies to clean energy and adapt to climate change. These organizations put the figure at a minimum $250 billion annually. Even that figure would be an almost unnoticeable redistribution of wealth from rich to poor, hardly enough to make a dent in the world's $60-trillion economy. Since most of it would inevitably be in the form of the installation and operation of technologies and equipment from industrialized countries, the money would likely remain in the industrialized countries, just as most aid money is eventually recycled back to the rich, who supply the goods and services.

In the dying hours of the convention, many poor-country delegates warned of the dire consequences of what they described as a shabby deal imposed on them by a backroom oligarchy of the elite economies. None was as direct as Lumumba, and his words caused indignation among Western countries. He leaned into his thin, goose-necked microphone with its little red light indicating he had the floor and got to work producing a hard display of anger and rhetoric, coolly charging that the Western democracies are content to march Africans to the climate change equivalent of the gas chambers. "This is asking Africa to sign a suicide pact, an incineration pact, in order to maintain the economic dependence of a few countries. It [the Copenhagen Accord] is a solution based on the same values that funneled six million people in Europe into furnaces."

Shocked Western delegates rushed into an orgy of righteous anger. One by one they lined up to protest against the charge leveled by Lumumba. Canada was among the first to pounce. A country that had done its best not to be noticed, like a student who hadn't done his homework, suddenly found its voice and joined the dash to the microphone to condemn Lumumba's rhetoric. Canada's chief negotiator, Michael Martin, claimed that he was outraged at the reference to the Holocaust and demanded its retraction. Sweden's chief negotiator, Anders Turesson, took his turn: "The reference to the Holocaust is, in this context, absolutely despicable." Great Britain's environment minister, Ed Miliband, called it a "disgusting comparison," protesting, "This is a document produced in good faith that is by no means perfect but which will improve the lives of millions of people. The other choice is what Ambassador Lumumba offers us. It is a choice of disgusting comparisons to the Holocaust and of wrecking this conference. What will the world think of us if we come out after two years with simply an information document?"

Several other Western nations chimed in with similar expressions of outrage. Nowhere was there any acknowledgment of the fact that in Africa today, climate change has created a cycle of drought, flood, disease and armed conflict that has brought whole populations to their knees, and that there was little in the Copenhagen Accord that would reverse so dire a situation. Forgotten by Canada, it seemed, was its unfettered tar sands development that ensured the continuation of a lifestyle that brought us to Copenhagen in the first place and whose soaring emissions are helping to melt the Arctic and destroy what's left of the indigenous Cree, Dene and Inuit tribal cultures in a sort of off-the-books ethnic cleansing. All the rich countries' delegates could see in Lumumba was a shrewd and manipulative servant of oil, of an alleged war criminal and of the Chinese who had just branded the industrialized countries—many of whom, like

Belgium, had been brutal African colonizers or slave traders—with the worst kind of blood libel. Lumumba withdrew into silence. But he did not back down.

Many G77 countries remained silent, but not the ALBA countries (Alianza Bolivariana para los Pueblos de Nuestra América): Bolivia, Cuba, Dominica, Honduras, Nicaragua, Ecuador and Venezuela. ALBA is a Latin American trading bloc set up to counter the United States' influence in the region. ALBA countries claim capitalism is the root cause of climate change and that the fight against climate change should include a massive redistribution of wealth. Two days earlier, both Bolivian president Evo Morales and Venezuelan president Hugo Chávez had warned the plenary about "top secret" backroom deals put together by the "world imperial dictatorship" of industrialized capitalist nations. The Accord met that description. Both men had left Copenhagen the previous evening after instructing their ambassadors to defeat the deal. The Europeans were well aware of this, but were helpless to stop it. "There is no way that Chávez could have allowed them to do that [vote for the Accord]," one senior European official said. "He would have lost face massively. So it was so tremendously predictable what happened on the Saturday. And we did predict it."

Rasmussen could have ignored the protests of the ALBA group as well as those of Sudan and declared a consensus. But as one European delegate said: "If you want to pull a stunt like this, you have to have respect, you have to have that kind of authority over the convention." And he clearly didn't—which is why the Europeans looked for a backup plan. They found it in a 1995 climate convention in Geneva where the parties simply voted to note the existence of an accord without giving it any legally binding significance. When British environment minister Ed Miliband saw there was no hope of a consensus, he moved that the convention note the existence of the Accord. The parties agreed. The exact phrase, which they placed at the top of the Accord's twelve paragraphs, reads:

> Decision -/CP.15
>   The Conference of the Parties takes note of the Copenhagen Accord of 18 December 2009.

The conference might just as well have taken note of the color of the president's pants. One month later, on January 18, 2010, Yvo de Boer sent a notification inviting all parties to state by January 31, 2010, whether they "wish to be associated with the Copenhagen Accord." De Boer stated that "in the light of the legal character of the Accord," the parties had only to send a letter to the secretariat stating their position. The United States and Canada quickly sent notes in the affirmative. At the same time, Canada used the opportunity to file with the UN substantially revised emission reduction targets: Canada reduced its target from 20 percent to 17 percent of 2005 levels, which amounts to a 2.5 percent increase over 1990 levels. Canada had now fully allied itself with the American target.

Other fossil fuel nations returned home to business as usual. Twitter followed the progress:

South Africa's SASOL wants to build mega (polluting) coal-to-oil plant in China http://bit.ly/1JbJlb (podcast) #350ppm #climate
1:02 AM Mar 10th from web

South Africa to ramp up thermal coal exports via new terminal at Richards Bay http://bit.ly/chyIY9 #350ppm #climate
1:05 AM Mar 10th from web

Massive demand for coal sends India looking to Africa & elsewhere for supplies http://bit.ly/b8VBZs #350ppm #climate
1:06 AM Mar 10th from web

Note: African National Congress party that Minister Hogan represents has huge profit stake in new coal power #350ppm #climate
2:24 PM Mar 13th from web

Australia's biggest-ever export deal: $60 billion (!) worth of coal, to China http://bit.ly/d96aFx #climate #350ppm
6:15 AM Feb 7th from web

Many countries pointed out to de Boer that the Accord had no legal standing. In response to the protests over the legality issue, de Boer sent out a second notification, on January 25, 2010, to clarify the first notification by acknowledging that the Accord provisions "do not have any legal standing within the UNFCCC process even if Parties decide to associate themselves with it." That prompted more than one hundred countries to associate themselves with the Accord. Ironically, four of the major creators of the Accord—Brazil, South Africa, India and China—all had lukewarm responses to de Boer's call for action. They sent in emission pledges but refused to ally themselves with their own Accord. Nevertheless, these countries called the talks a success. Indian environment minister Jairam Ramesh explained, "The reason why I said it was successful was because we protected our national interests. And for me I went to Copenhagen not to save the world. I went to Copenhagen to protect India's national interests. And my mandate was to protect India's right to foster economic growth."

That sparked Todd Stern to complain in a speech sponsored by the Center for American Progress, a think tank for which he once worked, that "the statements that we have seen from China and the other BASIC countries do evince a desire to limit the impact of the Accord." Stern claimed that reactions from many countries were "ambiguous" and amounted to "cherry picking" the parts of the Accord they liked. "This is not after all a casual agreement," he observed.

That, however, is exactly what it is: an orphan born of a one-night stand that has no legitimacy within the UNFCCC. The convention's legitimate children—the Kyoto Protocol and the Long-term Cooperative Action treaty—are being scorned by the West and nourished by the developing world. The United States, Canada and Australia have steadily increased their emissions since 1990 as their gross domestic product soared.[3] There exists a rare consensus among economists that reducing carbon emissions will slow economic growth by only about one percentage point at the most. Yet these countries have insisted over the past ten years that their unfettered economic growth trumps action on climate change.

Just as emerging nations see their chance for a better life, the West turns the tables and demands emission cutbacks that could stall their growth. The lifestyle gap between rich and poor countries is enormous and continues to widen. It's partially reflected in per capita emissions. India, for instance, emits about 1.3 metric tons of carbon per capita, while the figure for the United States is 19 tons. The world over, more than 1.3 billion people don't have access to electricity; this includes more than 400 million Indians. India wants to put a car in every driveway, just like the United States. How does that reconcile with the world's need to reduce emissions? At the UNFCCC climate talks in Bonn in March 2009, I put the question to India's lead negotiator, Shyam Saran. He said: "You cannot say that people in India, in developing countries, should not have aspirations for a higher standard of living. So you cannot say, 'You stay where you are because you are a latecomer and we get to stay where we are because we have had a higher standard of living for the last so many years.' That is simply not politically salable."

Saran's reply defines the basic political dilemma of trying to stir action in a civil society whose ambitions are geared to infinite economic growth because, as we have been taught, that's what puts food on the table, that's what brings success in life, that's what distinguishes between those who ride and those who are sidelined as mere pedestrians. Several months after Copenhagen, Kamal Nath, India's minister of roads and highways, traveled to Canada seeking investors and contractors to build 7,000 kilometers of roads each year until India's 3.2 million square kilometers are fully linked by four- and six-lane highways. Estimated cost: $74 billion.[4]

There is one major flaw in India's argument. India is a poor country because it has a relatively low rate of literacy (68 percent) and is overpopulated. Since 1990, its population has increased by 301 million people—the population of the United States. And it is still increasing. Why should the world be held to account for India's inability to control its population and educate its people? Why should the world sit back and allow India to build highways to the high life, burning the energy it needs to bring its 1.1 billion people close to the same economic level as industrialized nations, when it knows that those emissions will likely toss humanity over the cliff of catastrophic climate change? India has been a democracy since 1947. Its failure to control its population and increase its living standards is not the fault of industrialized nations. Yet India now asks them to pay a high price for India's ambitions. It was not an issue that Saran could address.

The same is true of China, population 1.3 billion and rising. The West is not responsible for China's six decades of murderous Communist Party rule that froze that nation in poverty while orchestrating a rape of the earth that is unprecedented in the history of mankind. The point is, there are no good guys in this story. We all ultimately have to be held accountable. No country gets a pass. At the same time, however, we all have a fundamental moral responsibility to look after each other and respect all life. We cannot turn to poorer countries and say, sorry, you are too late.

What the developing countries fear are trade sanctions. Obama's talk of protectionism worries China. Before Copenhagen, he slapped a 35 percent tariff on Chinese tires. A 36 to 99 percent tariff on steel pipes was pending, with the threat of more of the same looming—a direct challenge to China's world-beating US$1-trillion export economy. Talk of a trade war had crept into Chinese newspapers. After rumors spread in Copenhagen that China had worked out a secret deal with the United States, I happened to spot Qingtai Yu, who is China's special representative for climate change talks and a senior foreign affairs official, nip through a side door near the media hall followed by a few aides. I made chase. The door opened onto a loading bay, where I found Yu lighting up a cigarette. I asked him if China intended to support the Accord, as Obama had claimed. He took a drag on his cigarette and turned away. But one of his aides spoke for him: "If we don't have an agreement in Copenhagen, it will make the case for border taxes stronger. So he says in the interests of strengthening our hands to fight the forces of protectionism, we must have an agreement in Copenhagen." Yu then stubbed out his cigarette and left.

Throughout 2009, Yvo de Boer had tried harder than anyone to urge industrialized countries to be more ambitious in their emission targets in order to reach an agreement in Copenhagen. The Accord fell well short of what he had envisioned. He tried to put a brave face on it by announcing that the document would be the basis for a new round of negotiations beginning in April in Bonn. Yet even he had had enough. Copenhagen had been a debacle and he blamed it squarely on the Danes. In an astonishingly candid letter to his staff, the executive secretary of the UNFCCC stated that the Danes had destroyed whatever trust had been created between parties by attempting to impose their own text on the convention. "The Danish paper presented at an informal meeting a week before [Copenhagen] destroyed two years of effort in one fell swoop. All our attempts to prevent this paper happening failed. The meeting at which it was presented was unannounced and the paper unbalanced."

He said that the prevailing distrust rendered any attempt to advance the debate over the Kyoto Protocol and the convention on Long-term Cooperative Action a failure. He called the creation of the Accord "disorganized" and "ill-directed," adding that little attempt was made to sell the Accord to the G77. "When it reached the plenary, attempts to get regional groups to discuss it failed. Complete chaos resulted. Here we reached the very edge of the abyss."

Finally, de Boer wondered if this spelled the end of the UNFCCC process. He claimed that inviting the national leaders "backfired" because it did not create the spark for an agreement. Instead, "the process became paralyzed. Rumor and intrigue took over."

Blaming the Danes would become a professional sport played over the next twelve months in the climate talks arena. It served to divert attention from the main issue, which was the refusal of both industrialized and emerging nations to reduce their emissions significantly enough to avoid runaway climate change. A little more diplomatic dexterity would have avoided the chaos and ill-will that characterized Copenhagen, but not the outcome. The failure that was the Copenhagen Accord was a product of something much bigger and far more complex.

De Boer announced his resignation from the UNFCCC on February 18, 2010. It was a total surrender. He was joining an accounting firm.

CHAPTER 4

THE FAILING GIANT

IN WHICH WE PROBE DEEP INTO THE AILING INCUBATOR OF ICE AGES

Voices from Cyberspace:

ipf meiklejohn:
These hysterical "warmists" have exaggerated so much that very very few people believe them—and quite rightly so.
7:54 AM November 13, 2009 from web

john marsh:
How did green land get it's name? Oh yes that's right it was covered in vegetation not ice. Was London flooded then? err NO was the sea level higher? err NO . . . this whole effort is an exercise in eugenics and taxation increase.[1]
4:25 PM November 15, 2009 from web

BedfordFalls:
The planet is in the grip of bad science and shockingly inept and gullible scientists.
8:11 PM March 10, 2009 from web

As I write this, the total parts per million of carbon dioxide in the atmosphere is 387. That represents a 38 percent increase over the measurement at the beginning of the industrial age—280 ppm, which is considered the norm for the human species. One reason scientists know that this increase comes from humans burning fossil fuels is that, unlike other atmospheric carbon dioxide, carbon dioxide from fossil fuels contains none of the radioactive isotope carbon-14.

Most of the growth in fossil fuel carbon dioxide has come in the last fifty years. When the Intergovernmental Panel on Climate Change began its work in 1989, the increase was one part per million each year. Now it's more than two. It is a scientific fact that the greenhouse gases blanketing the earth's atmosphere— carbon dioxide, water vapor, methane, nitrous oxide, ozone and chlorofluorocarbons—are responsible for maintaining a temperature range that makes life on Earth possible. Venus and Mars also have greenhouse gas atmospheres. Venus's is very dense, which is one reason that planet's temperatures reach 467 degrees Celsius and it is now surrounded in sulfuric acid clouds. Scientists theorize that four billion years ago the surface of Venus had the same water characteristics as Earth but, for reasons that remain unexplained, the water evaporated. Mars's atmosphere is thin, which is part of the reason the planet's mean temperature sinks to as low as minus-143 degrees Celsius and the planet is freeze-dried. Its water either gassed out billions of years ago or froze into permafrost that is several kilometers thick.

In between the two is Earth. It exists in a storybook orbit that scientists refer to as the Goldilocks zone. Here temperature variations are not too hot, not too cold. Just right. The slightest change in heat input, however, can threaten that fairy-tale existence, just as a slight variance in body temperature can send a person to their bed.

An ice age is triggered roughly every hundred thousand years by a combination of tiny variations. These include a change in the shape of the earth's orbit; a slight tilt in the earth's rotational axis (about a degree and a half); and changes to the time of year at which the earth is closest to the sun. These changes alter the seasonal and latitudinal distribution of solar radiation received at the earth's surface. Essentially, the earth tilts slightly away from the sun, reducing the total energy received during the summer. It's called orbital forcing and the result is that not all the previous winter's snowfall melts at high latitudes. Over thousands of years the snow builds up. It eventually collapses under its own weight to form ice, and the ice begins to flow over the continent. This causes an increase in the albedo feedback, in which solar heat reflects off the expanding snow and ice surfaces instead of being absorbed by the earth. The process induces more cooling, which enables the ice to build up more quickly. During this period of glacial expansion, the amount of CO_2 in Earth's atmosphere drops eighty parts per million, or 30 percent. For the last four ice ages, the drop has been from 280 to 200 ppm. This drop allows more heat to escape back into space, reducing the greenhouse effect and cooling the earth further. [2]

This is the background against which modern global warming is occurring. Beginning about 900,000 years ago, there have been nine ice-age cycles. It takes about 90,000 years to get full glacial conditions. Then Earth's rotation swings back and we get another energy kick. The ice sheets are so large at that point, and extend so far south, that they are extremely fragile and collapse immediately—in the space of 10,000 years. Left behind is a giant ice-age scar across northern countries such as Canada. Then the seas warm. Vegetation spreads. Levels of greenhouse gas rise back to 280 ppm. (This is a paradox, because the expanding vegetation, which is substantial, actually sequesters carbon.) It's all about water transfer: ocean water to precipitation to glacier ice, and then back to the ocean as meltwater. Sea levels rise and fall with the glacial cycle. Scientists calculate that this rise and fall was about 120 meters in the last ice age.

A very slight change in the earth's orbit and axis of rotation triggers an ice-age cycle. The resultant drop in temperature at high altitudes starts the ball rolling. The reduction in atmospheric CO_2 content enhances the cooling. Scientists still don't completely understand the chemical process that would cause this drop in CO_2 levels, or why there is a correlation between CO_2 and ice volume. One likely factor is the dissolution of CO_2 in the cooling oceans, because colder water can absorb more CO_2 than warm water. In any case, climate models show that without that 80-ppm drop in atmospheric greenhouse gases, the ice ages are inexplicable. The metronomic ups and downs of CO_2 are critical to climate.

Each year, we emit more than 29 billion metric tons of greenhouse gases into the atmosphere from the burning of fossil fuels. Since the beginning of the industrial age in 1750, the CO_2 content in the atmosphere has climbed to 387 ppm—38 percent higher than the normal maximum interglacial level. The earth's atmosphere is thickening and the earth is getting warmer. Even if we stopped adding carbon dioxide, it would take several centuries to remove the excess from the atmosphere. The natural cycle of carbon transfer between plants and the atmosphere, which is the basis of life itself, has been put out of whack. But we are not stopping; 2010 broke another record in man-made emissions. The International Energy Agency estimates that by 2030 emissions from fossil fuel burning will increase another 38 percent, to 40.2 billion metric tons.

While there are still uncertainties about the long-term effects on climate, what we have seen so far is an irrefutable warming of our world by almost one degree Celsius, caused primarily by man-made emissions. History tells us that we should be entering a new ice age. Science tells us that over the last eight thousand years, solar radiation has been declining as the Arctic slowly tips away from the sun. That decline in radiation is still going on. We should be seeing the beginnings of a buildup of Arctic glaciers on Canada's Ellesmere Island, on Baffin Island and across the Queen Elizabeth Islands. The Arctic and mountain glaciers that have hibernated since the last ice age should be reawakening to begin their reconquest of the northern hemisphere. We should be seeing ice nucleating in these high continental latitudes and altitudes, which are the regions that incubated the giant glaciers that drove ice ages over the last 2.5 million years and shaped the landscape of almost all of Canada, Greenland, northern Europe and Russia.

Instead, what we are seeing is massive melting of the remaining ice-age glaciers such as the Greenland ice sheet, which has never melted since it was formed millions of years ago. We are seeing seas warming and losing their ability to absorb the extra carbon we put into the atmosphere. We are seeing permafrost thawing and releasing increasing amounts of methane. We are seeing species disappearing because they cannot adjust quickly enough to the changing climatic conditions. We are heading back 55.5 million years to conditions similar to those at the Paleocene-Eocene Thermal Maximum, when an increase in greenhouse gases comparable to what we are experiencing now—but much slower in pace—warmed the earth an estimated six degrees Celsius. That warming took twenty thousand years and led to the mass extinction of species. At our present rate of emissions, we could achieve such temperature increases in a century and a half. You could say that global warming is a good thing because it will save us from the scheduled ice age. But the downside is equally serious. If we fry, this is a bad thing.

Just above the earth's atmosphere, at an altitude of about 500 kilometers, two chunks of metal are following each other on a polar orbit. Flying at a speed of about 480 kilometers per minute or eight kilometers per second, they take ninety minutes to circle the globe.[3] They are called the twin GRACE satellites: GRACE-1 and GRACE-2, a.k.a. Tom and Jerry, after the cat and mouse cartoon characters. GRACE is an acronym for Gravity Recovery and Climate Experiment. The 3.1-meter-long trapezoid satellites are identical in every way except for their ground and inter-satellite communication frequencies. They are starkly geometric, compact and sleek, a bit like floating stealth Kit Kat bars. The chunky variety.

Tom tracks Jerry around the earth along the exact same orbit, with about 230 kilometers separating them. They are attached—in a manner of speaking—by microwaves. Their main mission is to map and remap the earth's gravity field. They make a complete remapping every thirty days, allowing scientists to study how the gravity field is changing. Gravity is the force at which two masses attract one another. The denser the mass, the stronger the attraction, and vice versa.

The earth's mass is constantly changing, from the surface down to the core. Little things change it. You put up a building—that will change it. You dig a hole and that changes it too. A volcano blows, a glacier calves a million-ton iceberg. Change again. The earth is a hot ball of liquid rock with a thirty-kilometer-thick crust over a plastic-like, flexible mantel. Load one area and it will bend inward, displacing the liquid below.

The two satellites calculate and display these changes. The way it works is simple. As Tom and Jerry orbit the earth, the front one is pulled forward as it nears areas of denser mass and therefore greater gravitational pull, increasing the distance between the two satellites. The front satellite then slows as it straddles the mass, shortening the distance between the two. Scientists measure these distances using microwave bands and GPS. The degree of accuracy is about ten microns, a millionth of a meter—basically the width of a human hair. Such accurate measurement of the earth's changing gravitational field had never been carried out before the satellites were launched in 2002. The main interest for the German and American scientists who operate the satellites is the changing glacial mass in the Arctic and Antarctic. Is global warming melting the polar ice sheets? And if so, to what degree and how is this changing sea levels?

Tom and Jerry have plenty of company. Hundreds of orbiting satellites are taking various readings, including temperatures (at a rate of twenty thousand readings per second), soil moisture, sea level rise, ocean currents and salinity, the earth's magnetic field, global wind profiles, clouds and aerosols (suspended particles in the air such as those found in smog and ocean haze). They probe for data so we can better understand the complex dynamics of climate change and better assess the risk that global warming poses. The earth has become a sick patient with dozens of diagnostic probes circling around it trying to figure out how bad the problems are and what the prognosis might be.

One of those probes is a European satellite called Earth Explorer CryoSat-2, launched into Earth's orbit April 8, 2010. CryoSat-1 never made it. It suffered a launch failure in 2005 when the Russian booster rocket veered off track after takeoff. It crashed into the Arctic Ocean, an irony not entirely lost on the scientists who had designed it to measure changes in polar ice thickness. So they spent another $191 million to build a new one, which was launched on April 8, 2010, from the Baikonur Cosmodrome in Kazakhstan.

It's not as elegant as Tom and Jerry. Looks a bit like a 4.6-meter-long old canvas army tent that has magically sailed off into space to orbit at 717 kilometers above the earth. Its speed is 7 kilometers per second and it uses a 3-D system that emits a cone of radar pulses 250 meters wide at a rate of 50 microseconds to measure ice mass down to the smallest cube. Its mission is to find out how fast and by how much the continental glaciers and the sea ice in the Arctic and Antarctic are melting.

Right now, CryoSat-2 is just warming up. The scientists are giving it several months to stretch its muscles and calibrate its machinery. They hope it will relay data for at least three years. But for the moment, there are problems. The main issue—one that has plagued previous satellites monitoring ice thickness—is the accuracy of the satellite's radar readings. It has to do with the shortcomings of radar and the essence of snow.

This is where Martin Sharp comes in.

Sharp is a glaciologist, and he's trekked over pretty well every major glacier in the world. He says glaciers are a key—in fact *the* key—climate indicator. You can argue over short-term temperature fluctuations, ferreting out tiny discrepancies in data collection and analysis to prove which years were hotter—or colder—than others, but you cannot argue over the melting of a glacier. Its loss of mass is as irrefutable as a melting ice cube in the sun. Glacier melt not only alerts us to what is happening now, it also warns us of what is coming down the road. Glaciers are the means by which that coming happens. They raise and lower sea levels. They warm or cool the earth. They help shape its contours and dictate its climate. That's why the science of glaciology has become a major thread in the climate change debate. Probing the world's ice fields—once little more than a historical search into a force that has shaped our earth over millions of years—is now a critical study of what will likely shape our immediate future. The science itself has taken on an unmistakable urgency of purpose.

Central to the fear of rapid climate change is the question of the stability of glaciers, particularly the ice sheets of Greenland and Antarctica, but also the small ice fields and ice caps of the Canadian Arctic. Sharp has studied glaciers for almost thirty years. His work has taken him to the world's great glacier fields in

Iceland, Alaska, the Antarctic, Norway, the Alps and the Canadian High Arctic. The one hole in his glacial CV is the Himalayas. Now fifty-three, Sharp says that's a young man's game: "High altitudes—and I don't like working with dysentery."

The High Arctic is an improbable spot for anybody to want to spend too much time. But glaciologists are a breed apart. They make their living dancing with an ice devil that in his angriest moments will swallow you whole. The glaciers often crack into deep fissures and crevasses, particularly when they are squeezed into narrow troughs of rock or flow down steep sections of the underlying bedrock. In winter, fresh snow can disguise these crevasses, which can be wide and deep enough to swallow a man and his snowmobile. One minute you are driving across smooth, snow-covered ice and the next you are dropping tens of meters into a yawning gulf. If that happens, it is next to impossible to get out. You suffer compound fractures and then you die slowly as the ice closes in. The cold stops the bleeding, but your brain keeps working. You feel and watch your own death. Unless you are lucky enough to die first from trauma or hypothermia, a fast-moving glacier will slowly close over you. So it's best to tread carefully and avoid areas where crevasses form. Although that's not always possible.

Some of the most ornery glaciers are in Canada, which is why Sharp immigrated here from England in 1993. "The Canadian Arctic was almost completely ignored by the glacier community. So I thought it was a big opportunity to work up there." London-born and Oxbridge-educated, Sharp says his parents taught him a love of the outdoors and of mountains. "That's how you get into glaciers. You want to be in the mountains."

Now a professor in the University of Alberta's Earth and Atmospheric Sciences department, Sharp began his work on the John Evans Glacier and the Prince of Wales ice field[4] on south-eastern Ellesmere Island, the most northerly island in the

Canadian Arctic archipelago. The costs of getting into and out of this remote region doubled over the years and he was forced to seek out a more accessible and less expensive research site. He was also looking for something a little less dangerous than the unpredictable glaciers of Ellesmere. He chose a perfectly shaped ice cap that covers the eastern half of the world's largest uninhabited island, shaped like an old sock and about 600 kilometers long and 160 kilometers at its widest (about the size of England). Located just south of Ellesmere, it's called Devon Island.

Devon's ancient ice cap closely resembles the giant ice sheet that covers most of Greenland. Spilling out to the north, east and south are huge outlet glaciers that feed Devon's ice into the sea. The ice cap is 12,050 square kilometers—small enough to allow scientists to study the whole thing. Relative ease of navigation reduces the difficulty of working on the ice. It's also one of the few ice caps that has a solid historical record of ice temperatures, velocity and melting, or what glaciologists refer to as "mass balance," which gauges the ability of a glacier to replenish

The Devon Ice Cap: This is what most of Canada would have looked like at the height of the last ice age. (COURTESY WILLIAM MARSDEN)

annually with snowfall what it loses through melting or the calving of icebergs into the sea.

We owe this historical record to a glaciologist named Roy (Fritz) Koerner.[5] In the annals of glaciology, this man is a legend. In the breadth of his annual Arctic journeys, he remains unrivalled. Beginning in 1961, Koerner, who like Sharp came to Canada from Britain in search of Arctic adventure and glaciers, began annual visits to the Devon ice cap for the Geological Survey of Canada. His first expedition, leading a team of four men, lasted a year. It was sparked partially by the Cold War and partially by Canada's concern that the Soviets and the Americans were challenging its sovereignty by running nuclear submarines in Canadian Arctic waters.[6] Canada needed to establish a presence and a better understanding of the territory. Above all, it needed to map the topography. The lion's share of the original funds for Arctic research came from defense agencies in both Canada and the U.S.A., as well as an eclectic assortment of private contributors: the tractor company Massey-Ferguson Ltd., the Hudson's Bay Company, the U.S. Weather Bureau, the U.S. Steel Foundation, along with some anonymous private donors.

Over the next forty-eight years, Koerner and his colleagues amassed a library of data on air and ice temperatures, mass balance and ice core samples that allowed them to study the climate record reaching back twenty thousand years. Koerner's research originally had nothing to do with climate change; it was the pure science of glacial-climatic interactions. Only later did scientists realize that in an era of rapid man-made warming, the response of the glaciers could have profound significance. The melting of glaciers and ice caps is now considered the dominant contributor to sea level rise. About five years ago it was number two behind warming oceans (ocean thermal expansion). But the increase in glacier melting together with the decline in thermal expansion due to the cooling of the Pacific Ocean, which in turn is due to more frequent La Niña conditions in recent years, has now made glaciers number one.

The data that Koerner collected is critical to creating a reliable historical record that will tell us if the Devon ice cap is now melting faster than in the previous fifty years and, in relative historical terms, how fast it is melting. Koerner studied Devon pretty well every year until his death in 2008. In fact, he missed only one annual visit, in 1969, when he took time off to ski across the Arctic Ocean from Alaska through the North Pole to Norway. He and his small party were the first to do that.

Another reason Sharp chose Devon had to do with his own peace of mind. He figured that spending weeks, sometimes months, camping on and sounding out the Devon ice cap and outlet glaciers would be a bit safer than doing similar work on Ellesmere, whose glaciers flow about twice as fast as Devon's and therefore have intense crevassing. Parts of Devon pose the same problem, but it's easy to avoid these areas if you choose. Also, because Devon is closer to Resolute Bay, the Canadian High Arctic transport hub, it is easier to mount a rescue.

I contacted Sharp in early April 2010, in the hope that he would be heading up to the High Arctic and would agree to take me along. To me, the work of glaciologists is fundamental to understanding climate change and particularly what is happening to our oceans. I had tried to persuade others to let me tag along on their field studies, but they hemmed and hawed and generally were reluctant to have to look after another body while they traveled across dangerous ice in an unfriendly climate. But not Sharp.

Sharp was a breath of fresh air. He was heading up to Devon Island not only to continue his studies of the ice cap's mass balance, its velocities and surface snow accumulation, but also to carry out experiments designed to help verify the accuracy of the data

CryoSat-2 would send back to Earth.[7] "I'm leaving in a week," he told me. "If you can get to Devon Island, you're welcome." I asked him a few basic questions about food (he would supply that) and weather (be prepared for minus-30 Celsius at least). Then he gave me a number to call at the polar logistics center to organize a plane or helicopter to his Devon camp and wished me luck.

CryoSat-2 is designed to take measurements in areas with complex topography. Glaciers often have mountains sticking up through the ice surface and steep slopes near their edges where they flow out to sea and where most of the melting and ice loss takes place. Previous radar-based satellites were designed to measure flat surfaces such as oceans and ice caps, sheets or fields, but were unable to obtain accurate readings around the edges of glaciers. Where the surfaces varied, the readings were biased towards the highest point, which could be a mountain that had nothing to do with the glacier. Laser readings from NASA's IceSat satellite, which burned out in 2009 after six years of operation, or other airborne lasers have produced only a spotty sample of data from those parts of the ice sheets where melting is most rapid. Their flight tracks often miss whole sections of a glacier, leaving scientists to extrapolate rates of change across the entire glacier. Some scientists have used this data to make fairly broad assertions about glacial melting. Others have been more cautious. Most of the caution comes from field scientists such as Sharp, who have a first-hand knowledge of what's on the ground. Nobody doubts the glaciers are melting; the question is by how much and what part of that melt is caused by global warming. Sharp hopes that CryoSat-2, combined with his ground work, will supply the missing data.

There is another, more fundamental problem that plagues radar satellites such as CryoSat-2. Rather than just reflecting off the ice cap surface, radar energy can penetrate the ice. As the radar signal penetrates the glacier, it will send back a signal each time the density changes. Ideally, this should allow scientists to measure the geometry of the glacier, including its height, and to track how it is changing over the years. The problem is that the radar signal is a giant cone-like waveform that sweeps over the glacier. Because the surface of the glacier is uneven, a whole host of different signals is sent back. The signal may confuse an interior layer with the surface layer one year but get it right the next, all in the same spot, jamming any idea of whether the glacier is expanding or contracting. Glacier melting due to climate change can amount to a few millimeters per year, while the radar error can be in the tens of millimeters.

Each year, the ice cap accumulates another layer of snow. With each annual accumulation, the snow underneath compacts under the weight of the snow above and becomes increasingly dense. Eventually it turns into what scientists call "firn," a precursor of glacial ice. Ice has a density of about 900 kilograms per cubic meter. The density of new snow is about 300 kilograms per cubic meter, and the transition from firn to ice takes place at about 810. So anything between 300 and 810 is being compacted to firn and eventually to ice. The radar signal responds to variations in density by reflecting back a stronger or weaker signal. The greater the difference in density between layers, the greater the chance of getting a reflection. The unprecedented summer melt since 1998 has created ice layers that are above the firn. This creates more confusion in detecting the precise layers indicated by the radar signal.

Sharp's job is to sort out the mess of signals so that false readings of the glacier's thickness don't slip through, rendering CryoSat-2 as much a pile of scientific space junk as its ill-fated predecessor.

Three weeks after the launch, Martin Sharp boards a plane in Edmonton and flies to Resolute Bay. He's on a five-week mission to make sure that CryoSat-2 is not a total waste of time and money. His ultimate destination is about twenty kilometers west of the summit of the Devon ice cap. He and his small team of three scientists leave Resolute Bay on April 26 in a twin-engine Otter. The aircraft is crammed with equipment of various sorts, two snowmobiles, tents and other Arctic camping gear. It also carries fuel, a shotgun (to fend off polar bears, although in the many years Sharp has been coming to the Arctic he has never seen one), and enough food to keep his team alive for much longer than their planned stay just in case they are marooned by bad weather, something that happens frequently in the Arctic, especially at this time of year.

Five days after Sharp had left Resolute for the Devon cap, I fly from Montréal to Yellowknife via Calgary. The capital of the Northwest Territories can be among the coldest places on Earth even in early spring, but when I get off the plane Yellowknife is positively balmy. People are wearing shorts. Runners trot along Ragged Ass Road, jogging by the bay. Several people are boldly kiteboarding over the Great Slave Lake's softening ice, which is already breaking up. Next morning at 4 a.m. I hitch a ride on a cargo flight that takes me 1,600 kilometers north to Resolute, which I reach about four hours later. Resolute is minus-14 Celsius. I pull my parka out of my duffel bag and grab my goose down gloves.

Resolute is the headquarters for the Polar Continental Shelf Project, a Government of Canada program that helps finance and supply logistics for scientific study and resource exploration in the Canadian Arctic. For years the government has talked about building a permanent military station and scientific research center in Resolute, but nothing has ever materialized. In fact, the government is cutting back its climate research money. Polar Shelf, which annually supplies ground and air support to about 130 scientific projects, has an annual budget of only $6.3 million, which has remained that size for years despite inflation. By comparison, the United States' annual Arctic science logistical support budget was US$89 million in 2007, US$90 million in 2008 and rose to US$103 million in 2009.[8]

I arrive in Resolute early on a Saturday morning under clear skies. It has changed a lot since I first flew up twenty-five years ago, landing at an airport that was little more than a dirt strip. Now it has a small but modern terminal, warehouses, a fuel tank farm, a new government logistics center and several private aircraft hangers, all designed to cater to scientists and an increasing number of Arctic tourists.

I check into the local bunkhouse, called the Narwal. It has a green, modular structure made of sheet metal and is managed by Atco, a Calgary company that specializes in installations and logistics for remote areas of the world. Its clients are primarily oil, gas and mining companies and the military.

The Narwal is where Arctic pilots, aircraft mechanics and various airport workers live during their tours in the region. Many work here during the northern hemisphere's spring-to-autumn season and then go south for six months to the Antarctic to catch the summer business of ferrying what the Arctic pilots call "mad scientists" to their fieldwork. The pilots and aircraft companies are all Canadian. They specialize in landing helicopters, Twin Otters— the true workhorses of the Arctic—Buffalos and refurbished DC-3s on snow and ice and glaciers in often dicey weather. Crashes are not uncommon and mechanics find themselves repairing badly beaten-up machines in the most remote areas and hostile climates of the world. At the Narwal, I meet a mechanic in the mess hall who has just returned from Melville Island in the western Canadian Arctic, where a Twin Otter had crashed on takeoff after hitting a rock in a shallow valley in a partial whiteout. The pilot had thought the ground was flat snow. The mechanic spent three weeks on the glacier as part of a team replacing the aircraft's ripped-out front wheel/snow ski assembly, as well as the nose, tail and bottom, plus electrics. Before that, he was in the Antarctic, where he'd spent a month replacing the props and other broken parts on a DC-3 that had crashed on a mountainside at an altitude of about ten thousand feet.

I spend most of the day hanging around the Polar Shelf office, where Tim McCagherty oversees flights for scientific expeditions. His walls are cluttered with maps, and one wall is taken up with an astonishing montage of satellite photos taken between March 14 and April 4, 2010, of the High Arctic ice above Ellesmere Island. The montage shows unprecedented fracturing of the ice stretching all the way up to the North Pole. The photos had been tacked to the wall by Trudy Wohlleben, the senior ice forecaster at the Canadian Ice Service. Several months later I telephoned her and she told me she had never seen so many fractures in the Arctic ice. She said she had put up the photos as a constant warning to pilots and scientists.[9]

At fifty-one years old, McCagherty has been working flight control in the Arctic since 1984, initially for several private companies and more recently for Polar Shelf. He is tall and wiry, with a large handlebar mustache, sparkling eyes and a silver ring in his left ear. His home is in the country town of Buckingham, near Ottawa, where his wife has a job with the Salvation Army. Through the whole "science" season, which ends in September, he works six-week tours, returning home for three weeks and then heading back to Resolute. Then they close down the station and he returns to Polar Shelf headquarters in Ottawa to prepare for the next season. Despite long hours and seven-day weeks, he maintains a cheerful voice over the radio that is often a welcome and reassuring diversion for scientists in the field. He radios every field party at 7:30 a.m. and again at 7:00 p.m. On the day I arrived, McCagherty had four parties on field trips in the High Arctic, including Martin Sharp's on Devon Island.

McCagherty tells me a helicopter is scheduled to fly out to Sharp's camp and I can hitch a ride. The weather decides when the copter flies and McCagherty has the weather patterns up on his computer screen. Low-hanging cloud cover is what you don't want;

it makes landing on snow too difficult. Its dark, shadowy representation on McCagherty's screen covers most of the western Arctic. I will be flying east, but the helicopter I'll be taking is marooned in heavy weather on Borden Island to the west. There, scientists with the Geological Survey of Canada have been using a remote control submersible to map the continental shelf to support Canada's offshore sovereignty claims. Bad weather has grounded the three helicopters that were helping them. The worst of it is the fog, which can roll over an Arctic island in minutes and which is not like southern fog, but rather is frozen crystallized water droplets that bite your exposed skin.

McCagherty spends a good chunk of the afternoon trying to find a place for a Twin Otter to land. Its destination has fogged over and Monica the pilot can't return to her point of departure because that too is now under fog. McCagherty marvels aloud at how fast the clouds have gathered. To make matters worse, Monica has skis—which the pilots call boards—on her plane and needs a strip of snowpack to land. That rules out Resolute or Eureka, whose runways are gravel. McCagherty finally finds a spot on Axel Heiberg Island, where McGill University has a scientific station. I spend pretty well the whole day watching weather patterns fluctuate across McCagherty's satellite screens.

That evening the weather on Borden clears and the helicopters return to Resolute.

I fly out the next afternoon with pilot James Hunt in a Bell 206 Long Ranger, operated by a company out of Goose Bay, Newfoundland. Hunt is from Montréal and was one of the pilots trapped for two weeks in the lousy weather on Borden Island. He's been a helicopter pilot for seven years, hoisting logs off mountainsides in British Columbia or ferrying government geologists and prospectors looking for diamonds and other mining resources on Baffin Island. He tells me he is getting bored with the Arctic, where he is too often stranded with nothing to do because of bad weather.

He hopes to fly in warmer places like Brazil. But when I tell him the rain forests can be as densely fogged out as the Arctic, he says he might rethink his plans.

The trip from Resolute to the summit of the Devon ice cap is about 380 kilometers, which should work out to slightly less than one tank of gas. Hunt keeps the Long Ranger at an altitude of about four hundred feet to stay beneath strong headwinds. We fly over the pack ice that still jams the channel between Resolute and Devon. Not a crack in sight. Just heavy blocks of angry snow-covered ice piling up against each other.

We cross onto Devon Island at 4:20 p.m. The western part of the island is flat and rocky, with drifting snow. If the planet keeps warming, it'll be a nice beach in about a century. Devon is largely cratered Cambrian rock, a sedimentary deposit formed more than 500 million years ago when the island was part of the continent. It is a rough sort of terrestrial analogue for Mars, which is why NASA and the Canadian Space Agency established the Mars Project on the island beside the Haughton meteorite crater. Scientists try to pretend they are on Mars testing out communications equipment, spacesuits, and manned and unmanned rover vehicles. They do their fieldwork in the summertime, stay in elaborate housing facilities, attract lots of media and leave before it gets cold, which is odd because Mars is a frozen planet. For Sharp and his crew, however, the summers have become too warm, making the melting ice cap and outlet glaciers of Devon far too unstable and dangerous for fieldwork. Their work now begins in late winter or early spring, when long nights have turned to long days and temperatures are still between minus-20 and minus-30 Celsius.

As we fly east, we approach a bank of frozen fog. Hunt goes higher to get above it and to see how far towards our destination it stretches. The engine roars and the blades beat harder. The fog isn't too bad, so we keep going. James's gyro isn't working—the dial keeps spinning around. He can't use a compass up here because it's too close to the magnetic pole, so he depends on GPS. The fog clears and we find ourselves approaching a miniature Grand Canyon, left behind by the ice cap when it receded. As soon as we hit it, the wind reaches up and grabs us, dragging the small helicopter down, tossing us like a toy. Hunt guns the engine. The copter shudders as it fights its way through. "It'll smooth out once we cross it," he says. And true to his word, the wind loosens its grip and retreats into the canyon when we reach the plateau. We are now crossing a broad plain of snow that could be mistaken for the ice cap, but it's not. Decades ago it separated from the cap and is now an orphan lump of stagnant ice covering about 1,960 square kilometers. With nowhere to go, it is melting away.

The ice cap itself is unmistakable. We soon see in the distance its defining border of brilliant blue ice, which runs like a smooth cliff over the barren land from horizon to horizon. If you core glacier ice in the Arctic, it comes out colorless and clear. After a while, it turns blue under your gaze as it begins to absorb and then emit colors of the light spectrum that are blue. The cap rises up behind the border in a perfect dome-like curve that fills the horizon. As we fly towards its summit, the cap soon outlines all 360 degrees around us for as far as the eye can see, a stark example of absolute space portrayed by a perfect circle of pure white. It is like a skullcap covering the top of the world.

We are 1,569 meters above sea level and still climbing the cap's gentle slope. After about twenty minutes I spot tiny specks of color in the distance. Red, blue, yellow, orange. Four tents. The air is so clean up here that colors radiate from a long way off. "There are the mad scientists," Hunt says. As we approach the camp, four people in bright blue and red parkas emerge to greet us. I remark to Hunt that there is something about the light up here that not only makes colors more brilliant but also changes them. The black of my parka has turned a dark purple. Purples turn to mauve. Clear ice becomes aqua blue.

A dusting of swirling snow curls up and around the machine as we set down. I wait for the copter blades to slow before alighting onto the hard-packed surface of one of the last remnants of the glaciers that incubated the last ice age and many more before it.

Martin Sharp and his colleagues stand well away from the machine, dressed in bundles of fur and bloated parkas and wearing boots the size of boats—a sort of informal welcoming committee. The air is frigid but dry, and with the brilliant sun it feels invigorating. Sharp towers over everybody else. A big bear of a man. We shake gloved hands and he introduces me to his doctoral student, Gabrielle Gascon of Montréal, David Burgess, an Ottawa-based glaciologist working for the Geological Survey of Canada, and his

(Left to right) Scientists Martin Sharp, Gabrielle Gascon, David Burgess and Tyler de Jong. We journeyed a good part of the day through heavy fog, but were greeted by a brilliant sun when we arrived at the Belcher Glacier. (COURTESY WILLIAM MARSDEN)

assistant, Tyler de Jong, a 21-year-old master's student at the University of Ottawa.

Hunt has taken off again to move his helicopter to the fuel dump at the other side of the camp. As soon as he finishes re-fueling, Burgess, Sharp and De Jong climb on board for a trip to check out the southern end of the ice cap. This ride I can't hitch on to because there's no room, what with all of their equipment. But their absence gives me a chance to get my bearings and feel the loneliness of a vast territory that has no natural markers and where compasses don't work. It is closing in on evening and the sun, surrounded by a giant halo, continues to track a steady circle around the horizon. An unreliable beacon. It would be a cinch to get disorientated and lost if you ventured too far.

It is mesmerizing to think that this silent beast of a glacier once expanded across the entire Arctic, joining other glaciers to invade the south, spreading over pretty well all of what is now Canada and parts of the northern United States, reaching a thickness of more than four kilometers and destroying everything in its path. The view I am looking at is what I would have seen 18,000 years ago standing on the future site of Winnipeg or Toronto or Montréal. It took the glacier 70,000 years to get there. Then it started melting. It took only 9,000 years to retreat to its Devon home, where it is now the largest of the three remaining ice caps in the Canadian Arctic. Scientists believe that about 7,000 to 10,000 years ago the Devon cap was considerably smaller than it is today and mean temperatures were about 1.5 degrees Celsius higher. This marked the warmest point of the current interglacial period, called the Holocene climate optimum. About 5000 BCE, the earth entered its cooling cycle and the glacier began to expand. It surged westward again during the so-called Little Ice Age, from 1400 to 1750—a time when the bubonic plague killed off about half of Europe's population.[10] Human populations stabilized and then increased, and the age of coal and oil brought forth the new fossil

Martin Sharp's tiny camp on the Devon ice cap. (COURTESY WILLIAM MARSDEN)

fuel economy. From 1750 onwards, greenhouse gas emissions grew and temperatures began to rise. The Devon ice cap retreated again to the eastern part of the island, where it now occupies 12,050 square kilometers, or 31 percent of the land mass—still big enough to reduce our group to six specks of dust.

An ice cap flows outwards towards its margins, under pressure from each additional layer of snow. The Devon ice cap spills into the ocean via outlet glaciers that have carved out valleys and canyons to the east, north and south of the glacier's summit. Where I am standing, the ice cap is about 500 meters thick. At its deepest, it is a little more than 880 meters thick. Were it to melt completely, it would contribute about ten millimeters to global sea level rise.[11] That doesn't sound like much, but Devon is a small player in a much bigger pond, and if the melting of this relatively tiny ice cap could have that effect, think what the melting of the massive ice sheets of Greenland and Antarctica could do. Greenland's ice sheet, first formed more than two million years ago, represents 2.8 million cubic kilometers of ice. Its melting would raise sea levels 7.2 meters—enough to wipe out most coastal communities.

———

Gascon shows Hunt and me to the mess tent, where we are to sleep alongside the Coleman stove, pots and pans, and a small diesel heater that, I am soon to discover, has a nasty habit of backing up and filling the tent with black smoke and carbon monoxide. The tent is about twelve feet long and six feet wide, held up by aluminum ribs and secured by long metal pegs jammed deep into the ice. Blizzards have built up snow around it, which paradoxically helps to secure the tent against the weather. I toss my duffel bag onto the thin plywood floor. The bag contains one thermal (minus-30 Celsius) sleeping bag, an inflatable groundsheet, one change of clothes, my cameras and laptop, and a survival kit including a utility knife and one eight-ounce flask of twelve-year-old Chivas Regal premium Scotch, which Sharp's equipment list indicated was an essential requirement for keeping sane while camping out on a glacier in freezing weather. And when the other three return from their helicopter journey to the south, we all sit around the heater on folding chairs and tuck into my whisky. When we polish that off, Sharp produces an enormous bottle of Irish whisky that miraculously lasts us through my stay. For water we simply melt snow. It comes out beautifully clear with no impurities and a drop goes just perfectly with the whisky.

That evening, we dine on spaghetti, meatballs and sausage, which Burgess had cooked up at his home in Ottawa and frozen inside a ziplock bag. Home-cooked frozen meals thawed and reheated in a frying pan over a portable propane stove are the staple of glacial camping. Sharp says good food helps sustain morale. Hunt pulls out a shaker of Kraft Parmesan grated cheese (the large size) and a bottle of hot sauce (also the large size), which he carries with him at all times in the Arctic as part of his personal survival kit; hot sauce and Parmesan will give anything taste. Gascon eats in silence. At about eleven o'clock, everybody clears out to their own tents. I spread my sleeping bag on the plywood then strip down to my long johns and crawl in. The bag is wonderfully warm even against the cold floor. But I soon discover a thin jet of snow blowing through a small gap in the bottom of the tent and across my face. I grab my parka, jam it against the gap and sleep through the night.

The next morning, I have to learn the procedure of glacier toiletry, and what's more I have to do it in a blizzard with wind so cold it freezes bare skin in minutes. Fact is, there isn't any procedure. You go outside, find a spot well away from the camp (but not so far that you will get lost in a whiteout) and, even more importantly, well away from the designated water supply (packed snow), dig a shallow hole, strip from the waist down and do the business hoping a polar bear doesn't saunter by, in which case you are toast. Then you pull your trousers back up, zip up your parka and cover up your offering. After a couple of goes you learn to build a wall against the wind and to carve yourself a small toilet out of the hard snow, one that allows you to prop yourself up, reducing the risk of falling. While it's never a pleasant experience, there is a sense of triumph at the end of it all. You have faced the Arctic's worst challenge and won. And you have discovered something about your anatomy that you would never otherwise have learned: the human bottom has amazing insulation properties.

Which leads to the question, where does your offering go? Glaciers are in constant motion, with gravity pushing the layers of ice down vertically towards the base and then horizontally over the base. Knowing exactly when something was deposited in a glacier helps scientists track the glacier's movement and velocity. The bodies of Alpine climbers trapped in crevasses occasionally prove useful because they give glaciologists an opportunity to plot a timeline from the day the climber fell to the day the glacier ejected the body. Macabre papers have been written on the glaciological applications of climbing accidents. Some scientists theorize that viruses land on glaciers and migrate with the glacial flow only to be ejected perhaps centuries later to begin another flu epidemic. My offerings will gradually disappear into the Devon ice cap and maybe a century or so later flow into the ocean via an iceberg calved from one of Devon's southern outlet glaciers. However, Sharp says the ice cap is changing so rapidly that it's hard to say when the glacier will eject the frozen little turd.

The blizzard grounds Hunt's helicopter and blows for most of the day. The Arctic is a desert. Its blizzards are mostly wind-blown snow picked up from one place and dumped in another. At the moment, the storm seems to be dumping all of the snow on and around our tents. There is simply nothing on this ice cap to stop it. Our tents become the blizzard's favorite toys.

Most of the day is spent digging out equipment and taking several core samples close to camp. Both Burgess and Sharp express surprise at the content of their cores, which reveal thick layers of clear ice near the surface. Such ice layers are an entirely recent phenomenon. Some of them are almost a foot thick and indicate a massive amount of summer melting at this spot, which is strange given its high altitude and cold summer temperatures.

Brushing off the snow gathering around the core sample to give it a close look, Sharp says that fifteen years ago he would have seen a wafer-thin layer of ice representing maybe a couple of days of summer melt. "Now we are getting enough days of melting that it is actually saturating quite a large thickness at the surface," he says. He pulls out a radar chart of the area that shows the different layers. Snow comes out various shades of red while ice is blue. "So we can really see there's a cold period here that's probably 2004, because that was really a cold summer. It was minus-17 Celsius here in July. If you have a cold period with snow, it will come out red because it is less dense. If you have a thawed period, it will come out blue because it freezes into ice." The blues are recent and dominant, indicating considerable melt over the past five years.

By the next morning, the blizzard has let up. Gascon, De Jong and I head out to collect data that will be used to help calibrate CryoSat-2. Burgess and Sharp erected a line of glacial monitoring equipment in 2004 that consists largely of metal poles with various sensors stuck in the ice. The line extends 48 kilometers south of the camp, from 1,800 meters down to 800 meters of elevation. It is identical to a line that the satellite will track when it is "science ready." The line was originally set up to help calibrate CryoSat-1. When that satellite crashed, Sharp and Burgess decided to keep up the annual readings. So now they have a clear record from readings taken over the last six years. Sharp uses his own ground radar, which penetrates about 7.5 meters below the surface—amounting to a snow accumulation of about twenty years. He says his ground-based measurements will help verify those taken by the remote satellite. By comparing the satellite readings to the ground radar readings and the core samples, scientists can edit out the "noise" inherent in the satellite's radar and detect exactly which layers the radar waves are reflecting off.

The advantage of a satellite is that it gives measurements of the mass of the entire Devon ice cap, which would be impossible for scientists to accomplish on the ground. The disadvantage is that the satellite's measurements of the true surface at any one point are not always accurate. Sharp's own ground radar locates the true surface, which then allows scientists to interpret accurately the satellite data. Other teams of scientists are doing similar work elsewhere in the Arctic.

After a breakfast of lumpy oatmeal, Gascon, De Jong and I leave to travel south down Sharp's satellite line. A fog of ice crystals has set in and Gascon uses GPS to guide us. She drives a snowmobile towing a komatik, an Inuit-designed wooden sled held together by rope lashings to give it flexibility over the hard snow and ice. We have loaded the sled with a wooden tripod we will use to mount a GPS and two metal boxes containing computers for storing the data. I sit on one of the boxes and cling to its sides as Gascon drives over the snow. Tyler drives the other snowmobile, also towing a komatik with another GPS plus equipment to measure snow depth and density. It takes us about two hours to drive to the end of Sharp's satellite line, following metal pole markers set about one kilometer apart.

At the end of the line, we stand the tripod on the sled, lash it down with rope and then screw the GPS antenna onto a fitting at the top of the tripod. Gascon switches on the radar and I sit on the komatik monitoring both screens. The GPS gives the exact position of the radar readings. We then begin our slow trip back up the line towards camp, sweeping the surface and subsurface with radar soundings, traveling at a snail's pace. At each marker, Gascon stops the snowmobile and I record the marker number, the time and the signal strength of the GPS and its satellite signal.

As we drive through a gamut of fog, blowing snow and clear skies, I tell Gascon I should get an honorary doctorate in glaciology for this work. She offers me a package of Cheezies instead, which I munch while taking readings and drinking green tea from my Thermos.

De Jong, meanwhile, is using GPS to determine the velocity of the glacier by measuring how the poles that mark the CryoSat-2 line have changed position since the previous year. He also measures the height of the snow relative to each pole. Burgess tells me later that measuring how the ice velocity changes over time helps show how melt affects the dynamics of different parts of the glacier, and how these dynamics contribute to changes in the ice cap's thickness. It's all part of an effort to understand how and why the ice cap's height and thickness are changing.

We return to camp late that afternoon. Fog still grounds Hunt, who has been hoping to fly east to the Belcher Glacier to help another team of scientists set up GPS and cameras to study the glacier's movement.

———

The next morning dawns with a clear blue sky. But a distant fog towards the Belcher Glacier unsettles Hunt, who decides to check it out. He lifts off and returns in about fifteen minutes, worried that if he goes to Belcher he won't be able to get back. For him, another day is lost.

Sharp, Burgess, De Jong, Gascon and I decide to take the three snowmobiles to the ice cap summit and then on to the head of the Belcher Glacier. Burgess has some coring to do along a line that reaches to the summit and Sharp wants to show me the Belcher, which he has been studying for five years. The return trip is about ninety kilometers. We set off through the icy fog.

David Burgess slides a sample of glacial ice and firn snow out of his coring drill. (COURTESY WILLIAM MARSDEN)

Along the way, Burgess stops to take core samples at every one hundred meters of elevation. De Jong digs out a square section of the snow about a meter deep. Burgess places his six-inch-diameter coring cylinder in the middle of the square, pulls the rip cord, and the two-stroke engine that drives the cylinder screams to life. The fiberglass cylinder has two metal chisels at the end that quickly carve out a perfect core a little more than one meter deep. Burgess pulls it out, spreads the core on the snow, adds another length of pipe to the drill and carves out another core. He repeats the action once more and in all comes up with a clean core length of about three meters. He measures, photographs and records the layers. Again, he and Sharp express concern at the thick layers of solid ice near the surface, particularly so close to the summit. At 1,920 meters above sea level, Burgess says it is hard to believe it gets warm enough in the summer to cause that much melt. All of his coring over the last few days has revealed the same extensive ice melt. He says that ten years ago you would not have found so much ice in a core here. Even at the summit, where temperatures used to remain below zero for all but a few days in the summer, there are signs of serious melting.

We pack up the equipment and, leaving behind the core, continue our journey. It takes more than an hour to get to the summit. As if by magic, clear skies greet our arrival. The top of the cap is pure white and sky blue from horizon to horizon. Crossed metal poles with a temperature recorder, wind speed indicators and other gadgets mark the summit. They are encrusted with feathered snowflakes. One sensor has been broken off and Sharp suspects a polar bear. Burgess disagrees. Not enough damage.

Ice doesn't move much laterally at the summit. It mostly sinks down towards the glacier's bed, where a kind of conveyor belt effect carries the excess mass downslope and over the underlying rock to fill the void left by the melting at the bottom.

We stand around as Burgess and De Jong drill their final

core. More measurements and photos are taken. This core also shows plenty of ice melt.

"The Devon cap summit was known as the pseudo-dry snow zone," Burgess says. "There are only two dry snow zones: the Antarctic and the Greenland ice fields. They are called dry snow zones because the temperatures never get warm enough for the snow cover to melt. Fritz Koerner coined the term 'pseudo' [for the Devon ice cap] because throughout his research he had seen melt only very rarely, like once every ten years from the early sixties up to the nineties. So in effect all you would get as an ice layer is maybe a two-millimeter-thick ring. What is really amazing to me—and I haven't drilled here since 2006—is to see an ice layer that is six inches thick. It shows that things have really changed a lot. If you can interpret that as part of a trend, a trend we are seeing in warming, it looks like air temperatures at high elevations tend to be warming or are warmer than they have been for several decades, for fifty years perhaps. Normally, and even in the time that I have spent up here—and I have been coming [to Devon] since the year 2000—it always seems much colder up here. It's the opposite of an oasis. It's a refrigerated area compared to the rest of the ice cap. But we don't feel that today."

Sharp says the ice layers are consistent with the climate data, indicating the summers have been extraordinarily warm in this part of the Arctic since 2005. "That means melt has spread to pretty much all elevations of the ice cap, which wasn't the case for at least the previous thirty-five years probably."

Probably? Sharp is cautious. He says the large layer of ice slowly turning aqua blue in the core at our feet is probably from the summer of 2007. "In the climate record, this was the warmest summer for the last fifty years for the Canadian Arctic, and in the mass balance record from the northwest part of the ice cap it's the most negative mass balance year.

"The other story . . . is this thing is showing up everywhere across the ice cap. And we also have this feature now in our radar data, which we didn't have in 2004, when we started making radar measurements. We now see this everywhere across the whole of the accumulation area that we have mapped, which covers the full elevation range [of the ice cap]." Normally, the whole process of snow turning to ice through compaction would take a century or two on Devon. Warmer temperatures have changed all that. Now snow simply melts into ice within a few summers.

What the ice core represents is not one summer of melt followed by dry snow summers. What has happened, Burgess says, is that in 2007 the snow accumulation melted and created a layer of ice. Subsequent years of snow accumulation partially melted and the water percolated down until it hit that dense ice layer. That layer then expanded. If this warming continues, the summit of the Devon ice cap could gradually turn into a giant ice rink, as have Arctic glaciers in Norway. It would then take on the characteristics of Devon's outlet glaciers, which are solid ice from top to bottom, changing the entire dynamics of the ice cap.

This melting creates yet another issue for CryoSat-2. The ice cap should be thinning as snow melts, drains down into the glacier and freezes. The satellite will record this thinning, but the data could be mistakenly interpreted as a reduction in the mass of the glacier. But the water isn't necessarily leaving the glacier—it is still there, in the form of ice. So a loss of ice does not necessarily mean that the glacier is disappearing.

"You measure the thinning and the natural tendency is to think that you have lost mass," Sharp says. "In fact, all you might be seeing is faster compaction." Coring helps validate the radar by clarifying the density of different layers, reducing the possibility of misinterpretation of the radar signals.

"We have to constantly validate throughout the life of the satellite," Burgess says.

———

We collect the coring gear, stash it in boxes, crank up the Ski-Doos and turn north towards the Belcher Glacier. In the distance are the smooth, rounded snow domes that define the mountains and hills bordering the head of the Belcher. I leave the track and drive towards the mountaintops to get a better look. Out of the corner of my eye I notice Burgess stop his Ski-Doo. I see him waving at me, but I keep going. I look back at De Jong, who is riding on the komatik behind me; he is also frantically waving and shouting. So finally I stop, flip open the mouth shield on my parka and yell over to him, "What's the problem?"

"Crevasses."

I had forgotten about them. I turn the Ski-Doo around and head back to the track laid out by Sharp, who is now a kilometer or so ahead of us. Burgess calmly reminds me that it's safer to stick to Sharp's track. My freewheeling days are over.

For about an hour we drive down the north gradient of the cap, following Sharp. Burgess is in front of me, towing Gascon on skis. Sharp turns east and we begin descending a steep and broad embankment of deeper, powdery snow, cutting across its side. My machine almost flips and I have to jump off to right it. I slow down and keep checking to see if De Jong is still on the komatik behind me. Each time I look back, he smiles and gives me the thumbs-up, and I admire his amazing balance.

Sharp and Burgess are about half a kilometer ahead, parked close to the edge of the Belcher Glacier. I pull up alongside them and the five of us trudge through fresh snow until we reach a snowy overhang that curls down over a cliff overlooking a deep mountain valley. The snow is so white the sunlight has turned each one of its tiny crystals into a prism of sparkling colors. We are standing on a carpet of diamonds.

Far below us are the telltale flow markings of the Belcher, pouring forth from the Devon ice cap in perfect stillness. We have a bird's-eye view of the sheer inclines where the ice cap feeds into the glacier on three different fronts. These are the headwaters, if you will, of the Belcher. It covers an area of about ten kilometers mostly to the south—near where I had gone independent. At the point where we are standing, the Belcher is about two and a half kilometers wide. On the far side is a ridge of small mountain peaks, each one encircled by glaciers. The distance from here to the terminus and the sea is about forty kilometers. Today there is no wind, no rifle crack of crevassing ice. Only silence. It is the quiet stillness of a million years or more of a glacier that knows nothing of the human element yet helps drive our climate and dictate our future. We stare at this muted monster without speaking and without knowing what is really happening. Knowing only warm sunlight and polar visions and, in the great distance, far to the east, the mountains of Greenland and the mother of all glaciers.

The Belcher extends about fifty kilometers into the ice cap and is the fastest of Devon's tidewater outlet glaciers, moving at a rate of up to three hundred meters per year. "Near the summit [of the ice cap], if we put a pole in and come back next year, it would have moved about ten meters," Sharp says. "Down here it would have moved probably by about a hundred meters. Further down even more. This is just like Greenland or Antarctica, where you have a large mass of slow-moving ice which is feeding ice into these glaciers, which are like rivers that take the ice out of the ice cap."

The glacier flows from an elevation of 1,855 meters down to the sea, where it ends in a towering cliff-wall of ice that defines a jagged coastline for eleven kilometers. Its terminus is below sea level. In fact, the last twenty-five kilometers of the glacier are below sea level. As sea levels rise, this creates a dangerous situation. "It [the glacier] will be stable until it floats," Sharp says. "Once it floats, the dynamics change completely. It could sweep out to sea, and that's what has been happening in Greenland. And we think it might be happening here too. We have this one site on eastern Ellesmere where the glacier doubled its velocity in

2002, and by 2005 it had thinned eighty meters at the terminus just by pushing icebergs out into the ocean. As it melts, the sea level rises."

Belcher calves the most icebergs into Baffin Bay of any Devon glacier—about 50 percent of the total. The main question facing glaciologists is, What does the long-term data indicate about the health of the Devon ice cap? At a time when it should be slowly expanding because of the onslaught of a new ice age, Devon is instead disappearing. The scientific data shows that Devon has declined steadily over the last fifty years, and that its decline has accelerated. Until recently, the altitude of the summit remained fairly stable, but the rate of thinning at lower altitudes towards the south has increased to a rate of about one meter a year, which is significant, Sharp says. Annual mass loss on Devon since 2004 is on average three times greater than over the previous forty-four years, owing to warmer temperatures. Burgess's transects show that the average loss during the last five years over all elevations is about thirty-three centimeters per year, compared with an average of only eleven centimeters during the last fifty years. Meltwater channels or streams have been forming over the last five years at higher and higher altitudes on the ice cap.

The melting and calving data do not tell the whole story, however. Standing in minus-20 Celsius staring down at the smooth and solid ice of the Belcher Glacier, you wouldn't think there's any melt at all. But what is happening underneath is what is important. Sharp's expertise lies in the drainage patterns of glaciers, how water plays and flows through, around and under them—a study that goes to the very fundamentals of glacier dynamics and climate change's effect on the ice. His first major research was on the Arolla Glacier in the Swiss Alps, a small and relatively unremarkable glacier that hugs the steep rock amid the Pennines. In the early 1970s it covered an area of 13.17 square kilometers. It has shrunk to about 6.5 square kilometers and thinned by forty meters in the last fifteen years. That's a lot of melt in a short period of time, and Sharp was there to analyze it.

Beginning in 1989, he spent seven years studying Arolla. He wanted to understand how glaciers switch their melting on and off and what happens to the water. Over time his team assembled a mountain of breakthrough data that changed the way scientists viewed glacial drainage. It also raised the possibility that the massive Greenland ice cap, and others like it in Canada and Antarctica, could quite suddenly—at least for an ice cap—slide into the sea. Not a happy prospect for the earth's coastal communities, where most of us live.

Sharp was able to show how water gets trapped and builds up underneath a melting glacier. The pressure from the water buildup literally lifts the glacier off its bed like a hydraulic jack. That's no small feat since a glacier's gigaton weight is enough to crush and bend the earth. "If you are in the right place at the right time, you can feel the glacier shaking as it lifts," he says.

The weight of snow accumulation in the upper part of a glacier pushes the ice down the mountainside. Most of the melting occurs at the bottom of the glacier. The ice moves down the gradient like a conveyor belt, filling the mass left by the melting. Only the friction on the rock bed stops a glacier from sliding all the way down to the bottom of the mountain. When a glacier gets beyond a certain thickness, it will start to deform under its own weight. Pressure from the top will cause it to surge downward.

"If you have a glacier which is entirely frozen to its bed . . . then the ice will slowly creep downhill under the influence of gravity and its own weight," Sharp says. "This changes once you raise the temperature of the bottom to the melting point. Then you have water under the glacier . . . and that water provides a lubricant. Once you have that, the glacier will continue to creep down the slope under its own weight, but it will also start to glide across the bed . . . Some glaciers go through surges where they can change velocity by a factor of ten." This glacial surge is what frightens glaciologists the most.

One question is whether this hydraulic dynamic is true of all glaciers. The answer Sharp found was yes. What is happening on Arolla is exactly what is happening on Greenland or on Devon Island on a much bigger scale. Sharp's models appear to explain why glaciers suddenly speed up. "It's not just a question of the glaciers melting more and it eventually disappears," he says. "When it melts more, you change the dynamics fundamentally. And that plays an important role in how rapidly the glaciers respond to climate change." So Greenland's massive ice sheet could slide into the ocean. Sharp says it's definitely a possibility.

Various other factors speed up the process even more. Geothermal heat will help melt the bottom of a glacier. This melting is enhanced by a glacier's insulation properties, which are similar to those of an igloo. Furthermore, friction from the accelerating glacier then adds more heat to the glacier bed, increasing melting and further lubricating the flow down the slope.

We return to camp by the same route we came. As we follow our tracks back towards the summit, I see a cotton whiteness across the horizon about twenty kilometers or so away; it's dense fog. We plunge into it just as we cross over the summit and turn east. Sharp is leading the way and his sled completely disappears. Burgess is next and I'm third. It's like entering another dimension. I can't see either Burgess or Sharp. I follow their tracks until I catch up to Burgess, whose snowmobile has stalled. He's pulling the rip cord but nothing is happening. He removes his goggles and gives me an exasperated look. He pulls the rip cord a couple more times and still the engine refuses to budge. "Did you check the fuel?" I ask.

"Ah," he says, slightly embarrassed. "No." He fetches the spare gas can and fills his tank. One pull and we are back on track.

The author at the end of a stormy day-long radar survey trip down a satellite line on Devon. (COURTESY WILLIAM MARSDEN)

A few minutes later we catch up to Sharp, who is patiently waiting in the fog.

An hour later, we are back in camp. Hunt has spent the whole day in his tent and he curses when we tell him about the fog-free Belcher. Next day he takes off for Belcher, where he stays past midnight ferrying cameras and GPS and other scientific equipment around the glacier.

After he leaves camp, another blizzard of heavy winds and blowing snow hits us. Burgess and De Jong are finished their fieldwork on Devon and are waiting for clear skies so a Twin Otter can fly them and their equipment north to the Agassiz Glacier near the top of Ellesmere Island. I will go with them. Sharp and Gascon are staying on Devon for a few more weeks. They need another stock of food and a new Ski-Doo.

We kill time by building an igloo. The trick is to find a quarry of hard-packed snow and then cut large blocks that can easily be trimmed to size. The quarry is not a problem. All the snow here is hard, so hard that you can pound it into position without it crumbling. What is a potential problem is that, under the first layer of cut snow blocks, we find a deep fissure in the glacier that is about six inches wide and runs right under my tent. The packed snow had disguised it. I consider moving house. But the thought is fleeting. It is too much trouble and I'm only here another day or so, depending on the weather. In any case, the fissure is so thin that I'd have to be a supermodel to fall in. But it is a good example of how the surface snow hides the fierce inner dangers of a glacier. The only indication that something is going on underneath the snow are the occasional rifle cracks of the splitting ice.

Our igloo construction takes most of the afternoon and part of the next morning. Two Inuit would have had it up in about one hour. It's big enough for only three people. But it's effective: not only does it insulate our body heat, it also acts as a sound barrier against the wind. Inside is total silence.

Our Devon home away from home. (COURTESY WILLIAM MARSDEN)

———

My final day on the cap dawns bright and sunny. The weather in Resolute has also cleared, as it has at the fuel cache[12] in Eureka. McCagherty radios that he's sent a Twin Otter for the trip to Ellesmere. Sharp and Gascon are packing up and moving their camp about thirty-five kilometers south. After a few days they plan to move again to the Belcher Glacier, where they will join two other scientists from the University of Alberta.

Our plane arrives after lunch, carrying food supplies and a new snowmobile for Martin; the clutch on his old machine sticks in high gear. Burgess and De Jong lift their snowmobile into the plane along with two komatiks, their tent, scientific equipment and personal gear, packing it to the rafters. The pilot, Lexi Larson, straps in the gear. We sit behind the cargo on foldout seats, our feet up on Burgess's metal boxes. Larson wastes no time. She seals the cargo doors and takes off.

CHAPTER 5

THE AGASSIZ VENTURE

IN WHICH WE DRILL BACK TO THE TSAR BOMB

As we fly north over Jones Sound and near the southern coast of Ellesmere, the toy buildings of the snow-covered settlement of Grise Fiord come into view through my starboard porthole. The village of 141 Inuit appears seemingly out of nowhere—two lines of tiny buildings neatly positioned along the shore. No roads lead into it. No roads lead out. It is hemmed in by mountain ridges and the frozen pack ice that angrily crawls up the shoreline each winter and then recedes in the cool summer months. I've been here before. I spot the mountain ridge we almost hit twenty-seven years ago when a powerful wind pushed our Twin Otter off course as it made its approach to Grise's airstrip. I can still hear the pilot cursing and yelling as he pulled the plane up and the Inuit women on board screaming, and I can see the mountains coming at us through the front window and then suddenly disappearing into a blue sky as the pilot muscled the plane left and over a ridge. He brought the plane around and attempted another landing, this time smashing down onto the frozen gravel runway as if it was all the plane's fault.

Leaving Grise to its splendid isolation, we fly over an ice field with a gaping crevasse down the middle, bordered by what appear to be soft lips of snow. Narrow canyons with glacial tongues cut through the ice plateaus. As we fly north towards the Prince of Wales ice field, canyons begin to round out into broader valleys. The land becomes a little less threatening. I envision it full of lakes, rivers, forests and wildlife. Climate change would look good up here.

We took off from Devon at about 1:30 and it's now 2:10. All of Jones Sound is frozen over. Not even a crack in the sea ice. You can see the island of Axel Heiberg to the west, frozen over with rolling cliffs against the sea.

The plane has limited heating in the cargo bay and it's cold. We are wrapped in our parkas. The snow from Devon Island that clings to our clothing doesn't melt. De Jong is lost in his earphones, listening to Rage Against the Machine's "No Shelter." Burgess tries to sleep through the whine of the aircraft. I stare out the window.

At about 2:50 we begin our descent into Eureka for refueling. As we near the runway, I spot a lone white Arctic wolf running over the rocks. De Jong is now sleeping, encased in his blue Canada Goose Expedition model parka. His Arctic boots, which have Sorel Intrepid Expedition labeled on the side and which are probably the biggest, bulkiest boots you'll ever see, are resting on top of a stainless steel box.

Eureka is wedged between Blacktop Ridge and the Slidre Fiord. In the distance, a cluster of modular buildings hug the shore-line. A Newfoundlander working in Resolute told me they have a good cook here—that is considered a major plus. To the west and across Eureka Sound, you can see Axel Heiberg Island. The rise and fall of the ocean and the expansion and retreat of the glaciers have traced their history over millions of years across the face of the coastal rock formations. This is true everywhere in the High Arctic, where every bit of its history is etched in stone.

The runway is permafrost gravel and the surrounding area resembles one big gravel pit, which at this time of year makes it hard to understand why the area is called the Garden of the Arctic. The reference, however, is to summer temperatures that can rise to over 20 degrees Celsius. Thirty kilometers east is Hot Weather Creek, one of the warmest places in the Canadian archipelago and getting warmer. In 1955, Jack McMillan, a geologist with the Geological Survey of Canada, discovered a fossil forest at Hot Weather Creek dating back 40 million years, to the Eocene era, when mean water temperatures around Ellesmere were 10 to 15 degrees Celsius and most of the earth was about 6 degrees Celsius warmer.[1] This was the Paleocene-Eocene Thermal Maximum, when the earth went through a very warm blip. Scientists believe it was caused by a slight warming that suddenly and catastrophically released methane from deposits of frozen methane hydrates (methane locked up inside an ice crystal) that lie at the bottom of the ocean. This sudden release of methane accelerated and increased the warming, causing massive climate change. Methane hydrate deposits become very unstable in warmer oceans. Scientists worry that our current warming could trigger another sudden methane release. It's not really known how much methane is in the oceans, but estimates are somewhere in the range of 20,000 trillion cubic feet—more than all other types of natural gas put together. Scientists are studying not only what caused this sudden warming but also how long it took the earth to recover. The current estimate is several hundred thousand years.

That warming blip was marked by a huge species extinction of deep-sea animals. But it also signaled the emergence of the modern mammal. It was a time when the High Arctic had lush forests of giant redwoods and deciduous trees, lakes and swamps, and a friendly, temperate and moist subtropical climate. A variety of reptile species, including turtles, crocodiles and mammals such as brontotheres, an early member of the rhino family, lived up here. Continents as we know them today had not fully formed. The polar islands did not exist but were part of the North American mainland, resting on the top like a dunce cap. Scientists once thought it was a local Arctic microclimate, but corings taken near the North Pole in 2004 showed evidence that this thermal maximum was more global. Yet microclimates do exist in the Arctic. Even today, the Hot Weather Creek area[2] puts the lie to low temperature readings at the nearby Eureka weather station.[3]

The talk at Eureka is all about the Twin Otter that plunged through the ice just north of Alert the day before. It was ferrying a team of U.S. scientists to the world's most northerly outpost, where they were going to measure the ice thickness. As the plane landed, the ice cracked and broke apart, and the hapless Otter sank beneath it, the scientists and two pilots scrambling out before the plane disappeared into the depths of the Arctic Ocean. I think of Trudy Wohlleben's satellite photos pinned to the wall of the operations center in Resolute. The plane sank through the thin ice at the exact time the *Geophysical Research Letters*, the journal of the American Geophysical Union, published a paper claiming the ice in this region was six meters thick. That would be enough ice to land a locomotive. The paper based its conclusions on data gathered by an electromagnetic induction sounding device towed across the Arctic Ocean in April 2009 by a DC-3. The authors concluded that the thickness of the old ice had remained constant and in some cases increased. But sea ice in the Arctic Ocean is generally incubated off the coast of Russia, where open water meets cold air, freezing the surface water into ice. Winds blow the ice across the Arctic Ocean, leaving behind open water where more ice is created. Eventually the ice gathers around the north end of the Canadian archipelago and Greenland, where it piles up into jagged castles of magnificent crystals. This pileup causes most of the thickness. Pack ice can sit right next to very thin ice, which is what the Twin Otter pilot found to his horror. Wohlleben's satellite pictures taken in early April 2010 showed cracked, thin ice throughout much of the region north of Alert. A scientific paper may well say the ice is six meters thick; the Twin Otter said otherwise.

The federal government established Eureka in 1947 as part of a network of Arctic weather stations. Since then it has expanded into an important scientific and communications research center in the High Arctic. One research endeavor is the Polar Environment Atmospheric Research Laboratory (PEARL), operated by a consortium of Canadian university research scientists.

A key element of PEARL's climate research is the energy exchange between the atmosphere and the surface: how clouds and aerosols affect the way in which energy is transported in the atmosphere and back. It's the key to understanding the exact inner workings of atmospheric pollutants, why Arctic ice is melting at an accelerated pace and why temperatures rise and fall in this freezing climate.

For that job, scientists bounce LiDAR (light detection and ranging) laser beams off particles and gases in the atmosphere to detect their location and nature. The thin green beam operates only at night, and you can see it shoot into the cold heavens against the starry sky. A telescope captures the return light as it deflects off particles. Scientists can detect a fairly full range of small particles such as CO_2, water droplets and ice crystals down to the electron level. The LiDAR also detects temperatures.

Most scientific research in the Arctic is done during the summer simply because it's easier. But what happens during the cold, dark months of winter is equally important. One of the basic precepts of climate change is that as the ice and snow melt in the Arctic, more heat is absorbed by the earth, which speeds up the warming process. This is simply because white surfaces reflect solar heat back to the atmosphere while dark surfaces absorb it. But what happens when there is no sunlight for four months, when both bare surfaces and snow have more or less the same dark complexion?

At night the snow, ice and bare surfaces radiate pretty much the same levels of infrared energy from the earth. So for four months in the winter there is no solar input but plenty of infrared radiation escaping into space from the natural heat of the earth. Because the Arctic is so cold, there is no evaporation. There is thus very little water vapor, which is the most efficient greenhouse gas, to reflect heat energy back to Earth. Heat easily escapes and the region cools like crazy. So you get an atmospheric condition in the winter called inversion. Normally, the higher up a mountain you climb, the cooler it gets. In an Arctic winter, it's the other way around; for about the first kilometer, the higher you go, the warmer it gets. The difference is not trivial. Eureka's temperature at sea level can be minus-40 Celsius and the temperature recorded by a PEARL lab at 600 meters is minus-20. How this plays on the ability of the atmosphere to trap pollution and to transport and transfer energy are things the Eureka scientists are working on. They are examining how each level of the atmosphere transmits, reflects and absorbs energy.

Greenhouse gas molecules all have three or more atoms arranged in a linear pattern so they can store energy by rotating or vibrating at particular frequencies. CO_2, for instance, has one carbon atom with an oxygen atom on each side. So when light at the right frequency zaps a CO_2 molecule, it will increase its energy level and the two oxygen atoms will move faster. The energy injection can take the atoms from a slow rotation—its ground state—into a faster rotation. Then one of two things can happen: the CO_2 molecule could spontaneously emit a photon—the basic unit of light—and return to its ground state, or it could collide with another molecule, which can also induce the emission of a photon. That photon could go anywhere—up, down, sideways. It could whack another molecule and start the whole process all over again, or shoot back to earth or up into space. In this sense the greenhouse effect is not, strictly speaking, the reflection of radiation off little particles in the atmosphere and back to Earth; it's an absorption and then an emission, and where the little photon will end up is anybody's guess. Molecules have no sense of direction. The only thing we know is that some photons radiate back to Earth. The more greenhouse gas molecules there are up in the atmosphere, the more energetic photons will be banging around up there, warming the atmosphere, and the greater the chance that more and more of them will come back to Earth.[4]

Understanding trends requires decades of data, and the scientists at Eureka say they need another ten years or so to comprehend what is happening in the Arctic atmosphere. But they probably won't get that much time. In January 2010, Canada's Conservative government shut down the Canadian Foundation for Climate and Atmosphere Sciences that for a decade has been funding climate research in Canada. All of the funding for the PEARL project comes from the foundation, which was set up by the previous Liberal government with a $110-million endowment. Consequently, the world's most complex remote atmospheric research station has enough money to continue operating only until the end of 2011. Maybe. Then the PEARL team will have to pack up their equipment and head home, hoping other scientists elsewhere will solve the mysteries of winter's Arctic atmosphere.

Meanwhile, some of the PEARL scientists are drifting off to another planet, preparing proposals to NASA to investigate the atmosphere around Mars.

After our pilot, Lexi Larson, has refueled and checked over her Twin Otter, she orders us into the aircraft and within no time we are back in the air heading northeast towards Agassiz. It's a brilliant day with an azure sky playing off against the white, white snow. You need strong sunglasses to shield against the blinding light. We fly over the Fosheim Peninsula and Canon Fiord. A little over an hour later, we sight the Agassiz ice field. This one is different from

Devon. It's interspersed with small saw-back mountains, their toothlike peaks sticking out of the summit of the glacier like the skeletal remains of a dinosaur. The glacier languishes around them before emptying into canyons and flat-bottomed valleys in the form of fingerlike outlet glaciers that stretch towards the sea. The Agassiz slopes much more steeply than Devon. It's smaller and more compact, about 16,000 square kilometers. The summit is about 1,850 meters above sea level—100 meters shy of Devon—but because of its slope it seems much higher.

We begin our descent at about 4:20 p.m. and Larson takes three tours of the summit before she attempts a landing. The hard snow whacks the boards and jolts the plane. Larson guns her engines and takes off again, not happy with what she sees. It's hard for a pilot to judge the snow and distance on so bright a day. There is not much definition. But if anybody can do it, Larson can.

One of many outlet glaciers that drain the Agassiz ice field, which rises in the background. (COURTESY WILLIAM MARSDEN)

A Saskatchewan native and veteran flier of both the Arctic and Antarctic, Larson has seen and experienced it all when it comes to polar flight. She pushes the Otter up along the slope and then veers off to the left and begins yet another turn around the summit. The hum of her engine increases, she touches down and then speeds up. She's testing the snow. She guns the engine and lifts off for another tour, taking us higher into the blue sky before circling back into the sun and then banking for the fifth time around. We're supposed to land about two kilometers from the weather station. As her landing boards hit the hard drifts, the plane rocks up and down and sideways. She throws her engines into reverse and comes to a sudden roaring stop. When the plane stabilizes, she starts taxiing over the ice field up the slope towards the spot where Burgess last year marked his coring with two metal pipes sticking out of the snow and ice. There she will dump Burgess and De Jong and all their equipment. In this white vastness she finds the two little galvanized posts without a problem. The wonders of GPS.

Larson climbs out of the plane and comes around to open the cargo door.

"Is this where you want to be?" She smiles, knowing she's dead on target.

"Yup. It's fine," Burgess says.

De Jong says: "Well, actually, I thought you were taking us to Ottawa."

In 1955, the Soviet Union began testing atomic bombs on the twin Arctic islands of Novaya Zemlya. Many tests were underground. Eighty-eight were atmospheric. The biggest blast came in 1961, when the Soviets exploded the fifty-megaton Tsar Bomb. The radiation fallout traveled right round the Arctic—and indeed the world—reaching its maximum in 1963. Snowfall deposited the radiation onto the Canadian archipelago, leaving a sizable trace layer across all the glaciers. That radioactive layer has since sunk. Even more than the dead bodies of ice climbers, radiation leaves an unmistakable time marker. By tracking the burial rate of that layer of fallout, Burgess can work out the net accumulation of snow and ice over time.

We pile out of the plane and begin unloading the various boxes, duffel bags, the tent, the Ski-Doo, the sleds and the gamma-ray spectrometer to measure the radioactivity level. Burgess and De Jong make camp next to the two poles. Burgess covered the hole where he drilled an ice core last year with a sheet of plywood. Now it's hidden by the winter's accumulation. He'll dig it out and send his detector down the same borehole to find the depth of the radiation layer and then the mass accumulation since 1963. This will allow him to calculate the long-term mass balance of the glacier—its net loss or gain.

This is one of four sites—the others are on Devon, Melville and Meighen, a tiny island west of Axel Heiberg—where scientists take annual measurements of the burial of the bomb layer.

Agassiz is ideal for the kind of deep coring that can provide a climate history dating back more than eleven thousand years. The nice thing, Burgess says, is that Agassiz is "well behaved." He means that it maintains consistent layers top to bottom for good coring. Because of pressures, stresses and flow patterns, a lot of glaciers jumble up their bottom layers. They kind of fold over each other in the same way a sheet of paper folds when you push it across a desk and it snags on something. Agassiz doesn't do that. Its layers have remained pretty consistent and well preserved right through the entire 10,000- to 11,000-year Holocene period. Our period. The only challenge is to account for the thinning of each layer caused by the flow of ice, particularly at the glacier's base, where it can be sheared off by the rocks. On Agassiz you can quickly spot anomalies such as the unusual summer melting Burgess has seen in recent layers dating back to the Tsar Bomb.

The other thing that's nice about Agassiz is that it's kept a fairly consistent elevation for more than ten thousand years. Every annual layer of snow accumulation has been established at about the same elevation and therefore is not contaminated by changes in altitude. This isn't always true of Greenland. "Some of the ice surface where Greenland cores were taken has changed by up to 600 meters in elevation," Burgess says. "I think in one case 800 meters. That introduces a big contaminating factor, if you will, in the climate signal. You don't really care what elevation it's at, but you want it to be consistent over the time period over which you are doing your interpretation."

Agassiz reveals that the world has warmed up about 1.5 degrees Celsius since the Little Ice Age and that we are traveling back six thousand years to when the world began to cool. "So we are on our way back up in a real big hurry compared to how long it took to cool things off," Burgess says.

Until recently, many glaciologists didn't believe a "warm period"[5] had affected the Arctic, producing temperatures considerably higher than today's. They claimed deep core samples of the Greenland ice sheet did not reveal signs of excessive warming. But Roy Kroener and a colleague at the Geological Survey of Canada, David Fisher, also a veteran High Arctic glaciologist, argued that historical variations in elevations of the Greenland ice sheet during the optimum interglacial period were masking a warm period. "The surface elevations [of these Greenland glaciers] had been dancing here and there through the Holocene," Fisher says. "The higher you go [in altitude] the colder it gets regardless of what the ambient temperature history is. So they were all masked—the ambient temperature records of a constant elevation were masked by the variation in elevation."

Which is why Agassiz is so important. Because its elevation has been relatively constant throughout the Holocene, there is no masking. Its cores clearly reveal a warm period.

One other glacier has had a fairly constant elevation throughout the Holocene: the Renland ice cap in eastern Greenland. Kroener and Fisher found that it too revealed a warm period. Still, scientists were slow to accept the fact that elevation differences had to be factored into the calculations.

"It's just sort of the way science lurches forward," Fisher says. "Roy Koerner and myself had been beating that drum for at least twenty-five years in saying that the central Greenland records weren't the right ones. But we're a very small band of two."

The so-called "warm period" holds such fascination for glaciologists because they believe it could help reveal the secrets of the stability of the Greenland ice sheet. Knowing what happened to it during the warm period, when the interglacial temperatures reached their highest point about 9,000 to 6,000 years ago, could improve our understanding of the effects on the glaciers of the rapid warming we are experiencing today.

In 2008, Canadian, Russian, Danish and French glaciologists set out to settle the "warm period" issue once and for all. They reassessed five deep-ice core records from Greenland and compared them with Agassiz's records. It's a bit like counting tree rings. Because Greenland sits between two major weather systems, it gets a sizable annual snowfall. So Greenland is good for dating cores.[6] Other ancient time markers include sulfuric acid from volcanic eruptions; there was a swarm of volcanic activity and earthquakes during the warm period. Some seismologists theorize that the melting of the glaciers stimulates shallow earthquakes and volcanic activity. You put a kilometer of ice on the surface and you will depress the earth by about a third of that amount, bending the earth's tectonic plates inward. When the ice melts, the glacial shift lightens the load on the earth's tectonic plates, opening space below for lava and liquid rock to flow. Tectonic faults get lubricated and move. The earth shakes. Volcanoes blow.

Seasonal changes can also be spotted in annual cores dating back as much as fifty thousand years in Greenland. Sixty centimeters of snow accumulation will be compressed to one centimeter of ice, but that's still enough to distinguish seasons, although that's about the limit.

Bubbles in the ice cores also are revealing of temperature. Ice formed from the compression of snow traps air bubbles. Ice formed by the refreezing of meltwater does not, and therefore is clear of bubbles. "In Agassiz, Roy and I spent a lot of time looking at these melt layers and they are an irrefutable measure of summer warmth," Fisher says. "Ice and snow melts at zero degrees. The more above zero, the more it melts. There is no argument over what it means." By "teasing out" the changes in altitude, as Fisher puts it, he and Koerner were able to show that the ice cores all reveal a "pronounced Holocene climatic optimum"—a warming— in Greenland during that period. After forty years of debate, the scientific community came around to the Koerner/Fisher way of thinking. "They have basically all signed on to it," Fisher says.

The warmer temperatures during the Holocene climate optimum—when temperatures were from 1.5 to as much as 3 degrees Celsius warmer than they are now—resulted in a loss of six hundred meters of elevation to major sections of the ice sheet. "That's a lot," Fisher says. "I worked it out and it's in the order of a meter and a half of sea level rise."

Fisher believes the margins of the ice sheet broke off and tumbled into the sea. The glacier then adjusted to the new reality. More ice flowed down to the margins and the elevations declined. "There is sort of a natural angle of repose [to a glacier] and it's pretty fixed," Fisher explains. "If you crack off a big percentage of the width or length, then the thickness has to change."

The new evidence showed that models had been far too conservative in their assessment of the impact of the warm period. "It is therefore entirely possible that a future temperature increase of a few degrees Celsius in Greenland will result in a Greenland ice sheet mass loss and contribution to sea level change larger than previously projected," Fisher and his fellow authors concluded in their paper.[7]

In all, sea levels rose 120 meters during the interglacial period due to the massive glacial retreat over North America and Scandinavia. If a few degrees of warmth can translate into an elevation loss of 600 meters to sections of one of the world's two last remaining ice sheets, the results of such a melt in modern times could indeed be catastrophic.

Whether the ice will simply slide into the sea is anybody's guess. But the science is clear that even in the frigid climate of polar regions such as the Canadian High Arctic or Greenland, slight changes in mean temperatures have caused significant melting of glaciers that has in turn led to sea level rise.

The cores from Agassiz essentially help glaciologists put in perspective what has happened to temperatures over the last 100 years compared with the previous 11,000. Ice cores from Devon give a good picture of what has happened over the last 60 years and particularly over the last 10, during which warming has greatly accelerated. They all show substantial and accelerated melting. "And we are on track to warm up over the next 100 years to what they were 11,000 years ago," Burgess says.

Scientists have been trying for decades to journey back even further in the Arctic climate record to the warming period that predated the one through which we are now living. This will take them back 115,000 to 130,000 years—to the so-called Eemian period.

Scientists know that during this last interglacial period the sea levels were much higher—about six meters higher—than they have ever been during the current interglacial. So one key question is how much of the sea level rise was attributable to the

Greenland ice sheet. Information about changes in ice surface height (and thus thickness) will help to address that question. It may also be possible to obtain information from the bed of the ice sheet that will elucidate how long the Greenland site has been covered by ice. Earlier ice coring in south Greenland led scientists to conclude that that site has been ice-covered for at least the last 450,000 years.

Scientists first drilled down to Eemian ice in southern Greenland in the 1970s, but they found that the ice was compressed to wafer-thin proportions, which made it unsuitable for reliable analysis. They tried again in the 1990s in central Greenland, but when they reached the bottom they discovered that the layers had folded over each other and jumbled up so much that the ice was too deformed to give a clean climate record. They tried yet again five years ago at a third spot in northern Greenland, but when they got to the bottom they found the ice was melting because of excessive geothermal heat flux. The deepest part of the historical record had vanished with the meltwater.

Yet these cores were not without tremendous significance. They pointed to abrupt "climatic flips"—about twenty-five in all—throughout the last glacier period. They also indicated that temperatures in Greenland during the Eemian interglacial were possibly about 5 degrees Celsius warmer than they are today. So understanding what happened to the glaciers during this period is critical to our understanding of what might happen as our carbon emissions force warmer temperatures. "Understanding the cause of these events, and their implications for future change, has become one of the hottest topics in climate studies, with significant policy implications," the Danish scientists said.[8]

As Burgess works on shallow cores on Agassiz, across the water in northwestern Greenland scientists are taking another stab at deep-coring into Eemian ice, where radio echo soundings reaching 80,000-year-old ice indicate there are undisturbed layers.[9] The $8.2-million venture involves primarily Denmark and the United States, with twelve other countries making relatively minor financial contributions.[10] They are drilling down 2,560 meters to the bedrock. They hope to finish in 2011 with an undisturbed ice core record of climate dating back 140,000 years that will confirm the temperatures of the Eemian interglacial period and what happened to the Greenland ice sheet during this warming. In total, the operation will have taken five years.

By June 27, 2009, they had cored 410 meters, to the snow that fell in 79 CE, when Mount Vesuvius erupted. A day later they hit the snow that fell in the year 1 CE. One month later they were into the transition from the glacier to the beginnings of the current interglacial period, 11,703 years ago—1,372 meters deep. One year later, in May 2010, the scientists were closing in on 2,000 meters. Another 300 meters or so and they expected to be in Eemian ice.[11] The deeper they go, the slower the drilling and the greater the danger that the ice will freeze around the coring equipment, choking it off. There is also an increased danger of the borehole collapsing before they reach bottom. So extra care has to be taken.

The need for greater historical understanding is becoming vital. Things are changing so fast in the Arctic in terms of ice melt and glacier velocity that scientists have to monitor the glaciers every year. "Our knowledge of what we need to understand has changed so dramatically in the last five years that we are really in a whole new ball game," Sharp says. "The science has been redefined in terms of what we have to do."

I ask him if he would be surprised if the glaciers suddenly did something totally opposite to what would be expected and reversed their decline as rapidly as they started it.

"Probably not," he says. "But I think the consensus is beginning to think more that we are going to hell in a handbasket and that there is no turning back. But we don't understand things well enough to know that that is true."

———

After Burgess and De Jong finish unloading their equipment, we part company. I feel guilty about leaving them behind. Burgess has been camping out on the ice for weeks and looks tired. But I have to fly to Europe and the renewed climate change talks. Bad weather is predicted and I can't afford to be trapped on an ice field for days on end.

Larson lifts off the Agassiz at about 5:15 p.m. and heads towards Eureka. I strain to see Burgess and De Jong, but they have disappeared into the whiteness. An hour and a half later we touch down in Eureka, where Larson again refuels the aircraft and then loads ten empty jet fuel barrels into the cargo hold before taking off for Resolute. The loud metallic crack of the barrel lids as they adjust to the altitude startles me. Larson takes no notice of it. We reach Resolute by 9 p.m. I grab some sleep and then get up at four in the morning to catch a flight to Iqaluit on the southern end of Baffin Island, which costs me $1,506.75. Another flight costing $881.08 gets me back to Montréal by the evening. It's overcast and about 8 degrees Celsius. My wife complains that it's cold.

As it turned out, Burgess and De Jong quickly completed their coring and were off Agassiz the next day. With heavy weather approaching, Larson got them out fast. They spent a night at Eureka before moving on to Resolute and then returned to Ottawa. I kept in touch with Burgess on and off over the next ten months while he prepared his data on the mass balance of the five Arctic glaciers he had visited in 2010, including Devon and Agassiz. On March 3, 2011, he sent me a graph plotting the ice mass balance of all five glaciers over the last fifty years. It shows a gradual decrease in ice mass from the base year of 1961 to about 1985, after which the loss accelerates. From 2005 to the present there is a precipitous drop that looks as though the plot lines simply lost their will to live.

Sharp and Gascon had a harder time of it. A massive blizzard that swept over Devon from the south snowed them in for four days, during which time they accomplished very little other than digging themselves out and running out of provisions. On the fifth day the weather cleared and they returned to the camp where I had left them and where there was a cache of food. They then moved to the Belcher Glacier, where they spent a sun-splashed week taking readings and installing more cameras.

The big Arctic melt begs the question of how the shift in load from the ice caps to the oceans balances itself out. If the ocean basins sink because of the new weight in water and the continents are rising because of the lost burden of the glaciers, does this offset the sea level rise? Is there a chance these forces will cancel each other out? For this I look to the Tom and Jerry satellites and scientist Richard Peltier.

It is June 24, 2010, and Tom and Jerry have been orbiting the earth for 3,021 days and eight hours. In that time they have completely mapped the earth's gravity field 97 times. This has allowed scientists to study the changes in gravity and therefore the changes in mass on Earth. Flows of surface and ground water, glacier melting, global sea level rise and disturbances deep inside the earth's crust can all be charted as Tom and Jerry trace and retrace their orbital patterns. The satellites have been sending back microwave and GPS readings since March 2002, when a Russian rocket shot them into space at the Plesetsk cosmodrome in the region of Archangel. So far, we have eight years of observations, twelve months a year. So what have they come up with?

Tom and Jerry are revealing a host of issues. Water tables in northern India and the western United States are losing water faster than they are being replenished. Most glaciers are melting. As the satellites have approached Greenland, the distance between them in successive months has been decreasing, meaning the ice sheet is losing mass. Studies of the data show that between 2000 and 2008, the Greenland ice sheet lost about 1,500 gigatons of mass, and GRACE indicates that the melting is accelerating; this water, of course, is ending up in the sea. Most of Greenland's melting is in the southern half of the glacier. The northern part as well as the interior have seen increases in precipitation, which means more snowfall and refreezing. But it's not balancing out. The Greenland ice sheet is melting on average 66 percent faster than it is replenishing itself.

That is not, however, the end of the story.

The added weight of the water melted and melting from glaciers is causing the ocean basins to sink towards the center of the earth. What's more, where the ice has melted, the land is pushing up. Eighteen thousand years after the North American glaciers began receding, the vast wilderness territories west of Hudson Bay in Canada, as well as the Arctic islands, are still rising as they rebound from being sat on by the last ice age. Most of Greenland itself has been squashed to below sea level by the weight of its obese ice sheet. The ice has crushed Greenland's central bedrock into a basin about three hundred meters below sea level. As the ice loses mass, the land underneath will begin to rise again. So when calculating sea level rises, this sinking/rising phenomenon has to be taken into account. The problem is how to make that global calculation.

Richard Peltier, a geophysicist at the University of Toronto, has struggled with that challenge for more than three decades. He is one of thousands of scientists around the world who seek meaning in this avalanche of satellite data. He's not a field scientist like Martin Sharp or David Burgess. He assembles data and then incorporates it into his mathematics and modeling work. Peltier is a small, cheerful man with a round body that looks as if it wouldn't meet the demands of the Arctic, but who knows. He has a suite of offices on campus and sits among his books, affable and ready to explain his world.

He has the habit of ending his sentences with "right?" or "okay?" or a combination of both. As in: "Jupiter is out there. Poseidon is out there. The moon is out there. Right? This tilt of the spin axis varies about a degree and a half, the consequence of the varied geometry in the overbody of the sun as a consequence of what we call gravitational N-body effects. Right? Okay?" This verbal tic often leads you to nod in agreement only to realize a moment later that you've missed the gist and need him to go back over it again, which you can tell slightly irritates him. But when you remind him that you are not a scientist, patience returns, an explanation follows and he's again off to the races.

An expert in large-scale planetary physics—the physics of the earth's interior and the evolution of atmospheres and oceans— he has modeled the earth's climate going back 600 million years. Some of his most recent papers analyze the data Tom and Jerry have sent back from space. "This is a very big deal, right? We are actually able to observe these polar accumulations of land ice to be diminishing, and the new mass in the ocean is pushing down just as land is pushing up as the ice is removed," he says. "So the ocean basins are sinking because mass is added. Over the past thirty-odd years I have developed a very detailed mathematical theory that allows me to describe all of these processes." He published his findings in 2009. "I made a prediction of what GRACE should see prior to the time the satellites flew, and it basically was confirmed by the satellites."

For Peltier, the budget is now balanced. Ice to water. The result is a net sea level rise of about 2.5 millimeters a year and counting. Greenland's contribution is running at about one-third of that. Were all of Greenland's ice to melt, it would raise sea levels seven meters (twenty feet). Goodbye London, New York, Miami, Venice, Tokyo, Amsterdam, most of lowland China, Bangladesh, Thailand, not to mention Tuvalu and most other Pacific islands.

None of this data went into the IPCC's fourth report (AR4), published in early 2007. The cutoff for papers that contributed directly to the AR4 was 2006. The first year of GRACE between 2002 and 2003 was not useful because the satellite system had not been properly calibrated. So, Peltier says, this data will have a big impact on the AR5. Unfortunately, it won't be published until 2014.

And Tom and Jerry are getting pretty long in the tooth, as Peltier puts it. They have maybe two years left before burnout. NASA and the German Aerospace Center hope to keep the twin satellites alive until 2015, but that's probably a bit of a stretch. There's a replacement satellite twosome on the drawing board. Instead of a microwave band joining the two satellites together, it will have a laser beam, which will give even more accurate gravity field readings than Tom and Jerry are able to do. Trouble is, it has no launch date and scientists worry about a data gap. The wider the gap, the more difficult it is to plot the models used to predict the long-term effects of climate change.

NASA has been using another satellite, ICESat-1—acronym for Ice, Cloud and land Elevation Satellite. It employs a laser to take altitude readings to track the changing contours of the earth. Since its launch on January 12, 2003, ICESat-1 has collected data on Arctic and Antarctic ice mass changes. It also tracks changes in the height of forest canopies to assess losses or gains in the earth's biomass. It is not, however, capable of collecting the deep-earth gravity readings that Tom and Jerry create. And its lasers burned out in October 2009, forcing NASA in the spring of 2010 to fly a DC-8 up to Greenland equipped with two lasers, radar, a high-resolution video camera and a gravity-measuring device plus a team of thirty-five scientists to take ice readings. Such a measure is expensive and cannot give the kind of steady stream of calibrated ice data necessary for building reliable climate models. But it's better than nothing and helps bridge the data gap until a new ICESat-2 is launched in 2015, or later.

There are more gaping holes in the global mass balance picture. There are entire regions with no data. The biggest is in Russia, where nobody is taking mass balance measurements. Scientists are reduced to guessing what is going on in the Russian Arctic. They use data from the closest regions, such as Norwegian glaciers, and extrapolate that onto Russia. But this often is not consistent with the summer temperature record in the Russian Arctic. In trying to understand the global situation, we are to a degree stuck in a Russian Gulag.

CHAPTER 6

AXIS OF MELT

IN WHICH INSANE MEN DREAM OF OIL, BIG GUNS, FLOATING
NUCLEAR REACTORS AND A NEW WORLD ORDER

MARCH 29, 2010. FIVE HUNDRED KILOMETERS BELOW TOM AND
Jerry on the 2,935th day of their mission, foreign ministers from
five Arctic coastal states—Canada, Russia, the United States,
Norway and Denmark—are meeting at a remote country estate
in the Gatineau Hills east of Ottawa.

The estate is about a half-hour drive from the capital. I'm in
a small bus along with other reporters from the Arctic Five countries plus a few Asian reporters from Japan and China, two countries that feel they have important interests in the Arctic. The
estate sits amid a mixed forest of tall pines, cedars, maples, birches
and ash trees and is part of the Gatineau National Park. In the fall
the forest is ablaze with reds and brilliant yellows. Now, in early
spring, the green cedars and pines stand out amidst a tangle of
naked hardwoods just beginning to reveal their buds.

As we pass three checkpoints along a winding road leading
to the main house, I spot RCMP sharpshooters lurking in the
forest. We stop at the coach house, where officers in black combat
outfits, pistols and bulletproof vests search through our bags and

make sure nothing explosive is hidden in our equipment. After that we are let off the bus. I lean against a stone wall listening to the officers chat about hockey.

Just up the road about fifty meters, the foreign ministers meet in a secluded, green-shingled Queen Anne–style house, the former retreat of a wealthy nineteenth-century inventor. It conjures up images of a bygone leisure class who, during the warm summer months, retired with their servants to the country to recline on spacious porches under laced gables and play in lakes whose pristine clarity seemed eternal.

Now the house is an empty shell stripped of its former glory and reduced to a trading floor where foreign ministers lay the ground rules for resource development in the Arctic. Stretched below it is the narrow expanse of Meech Lake. Usually at this time of year, the lake is still hidden under thick winter ice, which survives into mid-April or early May. But March has been strangely warm, with record-high temperatures of 14 degrees Celsius. For the first time in the meteorological history of Canada, the entire nation from east to west—excepting Newfoundland—is experiencing record-warm temperatures that meteorologists predict will last another three months at least. So the lake is rippling back to life much earlier than normal. Warmer water likely will enhance algae growth, which over the last ten years has helped transform many glacial lakes in eastern Canada into a muddy green slime that by midsummer makes them look like sewers. Some algae are so toxic that animals die when they drink the water. As climate change opens the Arctic for business, it is helping to destroy the familiar clear blue lakes of Canadian picture postcards.

The meeting is held in secret. The politicians enter among quick-moving processions of solemn aides and bodyguards, ignoring the Greek chorus of reporters corralled off to one side. The ministers say little about their mission. But make no mistake, they are here to lay the groundwork for a new world order—to set rules for policing the rich treasure trove of the Arctic and its priceless shipping lanes. The press has labeled them the Arctic G5. But really they are the Axis of Melt.

The meeting is unusual. The custom for Arctic countries is to meet as part of the Arctic Council, which was created by the Ottawa Declaration in 1996 to foster cooperation in the development and environmental protection of the Arctic. The Council's headquarters is on the tiny mountainous island of Tromso, Norway. Despite being 343 kilometers above the Arctic Circle, Tromso is warmed by the Gulf Stream and is a place of lush forests and verdant grazing land. The Council is supposed to be the central forum for all Arctic discussion. Indigenous people of the Arctic are participants. Observer status has been given to France, Germany, Poland, Spain, the Netherlands and the United Kingdom. Canada blocked observer status for the EU because of its ban on seal products. China has ad hoc status, which means it has to ask for permission to attend meetings, which it always does and always gets. It is seeking permanent observer status.

Now, however, the formation of the Arctic Five indicates a fresh intent to take command of the region. These are the only countries that have an Arctic coastline, which gives them a domineering presence. Their Arctic domain represents 6 percent of the earth's surface, of which about one-third is land, another third is continental shelf to a depth of five hundred meters and the rest is deep ocean. Under the retreating land and sea ice are petroleum and mineral resources estimated to be worth trillions of dollars. Geologists mainly from Canada, Australia and the United States each summer comb the vast open and now ice-free spaces of Greenland and Arctic Canada for diamonds, rubies, gold, and signs of the 90 billion barrels of oil and gas equivalent that the United States Geological Survey estimates is up there. Left out of this inner circle of five are the three other members of the Council: Sweden, Finland and Iceland (which isn't technically in the Arctic).

The Arctic Council has no legal status and the polar countries are happy to keep it that way—it is a forum for discussion and research. The Arctic Five is a forum for domination.

The recent flow of scientific data signaling the retreat of the Arctic glaciers and the melting of its sea ice dictates the necessity of the meeting; geography and the lure of wealth decide the players. Two decades ago the foreign ministers of these five countries would barely have given the Arctic a second thought. Now the data streams show the undeniable trend of temperature rise and glacier and sea-ice decline. About 6,000 ships navigate Arctic waters each year. Half of these ply ice-free waters just outside the Arctic Ocean off Norway, Alaska and western Russia. Roughly 1,600 are fishing vessels. The rest resupply Arctic communities, transport tourists or move ore from Arctic mines. Over the next twenty years, the Arctic Council reports that climate change will dramatically alter that shipping pattern. Arctic waters will become a ring-road highway for world trade. The sea ice that once jealously guarded the most remote areas of the world will open to the global fleet of about 25,000 oceangoing tankers, bulk carriers, containers, cruise ships and factory fishing vessels.

These countries know that climate change is rapidly reshaping global politics and economics. This is not simply about new shipping routes and the promise of oil and gas, diamonds and numerous other minerals. It's about the emergence of a new political order. As ice loosens its ownership of the planet's last frontier, the Arctic Five are preparing to snatch the deeds.

"The Arctic ocean region is on the verge of significant and fundamental changes," Canadian foreign minister Lawrence Cannon told the press after the day's session. "Today we discussed the emerging issues we all face in the region by virtue of our sovereignty, sovereign rights and jurisdiction in the Arctic

Ocean . . . We clearly understand that the potential of the north is a vast magnificent treasure we hold in trust . . . The natural resource potential of the Arctic Ocean is immense."[1]

The emergence of a new power bloc has not gone unnoticed by major exporters such as China and the European Union. Li Zhenfu of Dalian Maritime University in China wrote an extensive analysis of the situation for the Chinese government in 2009 and gave this warning: "Whoever has control over the Arctic routes will control the new passage of world economics and international strategies."[2]

Europe, China, Japan and India insist that, like Antarctica, its polar opposite belongs to everyone and the benefits of its exploitation should be universally shared.

Such a growing consensus among non-Arctic countries alarms the Arctic Five, and this meeting is intended to disabuse the rest of the world of its fantasy. A week earlier, President Dmitry Medvedev stated that outside interference in Russia's Arctic resources was "absolutely inadmissible." And two months earlier, Russian ambassador to NATO Dmitry Rogozin told the news channel Vesti-24, "The twenty-first century will see a fight for resources, and Russia should not be defeated in this fight . . . NATO has sensed where the wind comes from. It comes from the north." Some NATO countries have favored a role in the Arctic. But Canada's prime minister Stephen Harper joined with Russia in dismissing this idea out of hand, stating in 2009, "They don't belong."[3] Now Cannon makes clear on behalf of all five members that none of them will tolerate any meddling. He essentially throws down the gauntlet. "Many states and institutions that have historically not paid attention to the Arctic Ocean are now turning their attention northwards. Some have offered uninformed advice about the governance of our peoples, our lands and our waters . . . Now let me be perfectly clear, our government has never wavered from the principle that we will always defend the interests and the perspectives of people of the Arctic and this is a principle which I believe my colleagues share. This is why we act clearly and firmly."

U.S. secretary of state Hillary Clinton is only slightly more conciliatory. She tells the media, "Significant international discussions on Arctic issues should include those who have legitimate interests in the region. And I hope the Arctic will always showcase our ability to work together, not create new divisions." Her words are well chosen. Only those with "legitimate" interests in the Arctic are welcome. The question is, who will decide who has legitimate interests? The answer, of course, is the Arctic Five.

The assertions of sovereignty and the saber rattling over Arctic possessions are not only about staking out territory. They are also a message to those outside the Arctic rim that these five countries are serious about their sovereignty over the region. Climate change for the Arctic Five is an opportunity rather than a tragedy. It's also a psychological jolt. An ice-covered Arctic has for centuries been the imagined backyard of the A5, a natural extension of their landscape, a vast territory taken for granted. The sea that is emerging from under the melting ice cap has challenged that vision. Each year climate change forces them to redraw their maps and reassert their claims. International maritime laws that guarantee access to the open sea for all nations— including their fishing fleets and their oil rigs—are creeping into play, an assault on the self-image of the Arctic nations. Their sovereign authority over an iced-in area they have always assumed was their own is now under threat.

Their concerns are legitimate. A free-for-all in their own backyards would be unacceptable. The prospect of the world's fishing fleets raping the Arctic seas as they have so brutally raped the rest of the oceans is intolerable. The A5 are perfectly suited to rape their own seas, thank you very much, and have the history to prove it. Arctic waters already produce more than one-quarter of the world's fish catch, thanks mainly to the Arctic nations.

The Arctic Ocean is a nicely defined pond that facilitates sovereign control. Two of the five nations that border it are among the most powerful in the world. Access can easily be restricted. This is why each of the A5 has announced aggressive plans to build up Arctic military forces. Much of these plans still remains on paper. But that's inevitable. These nations follow the progress of climate change while probing global reaction and setting the parameters of the debate. Meanwhile, they organize joint Arctic military games.

Norway has purchased 48 F-35 Lightning II stealth fighter-bombers for delivery in 2014 and has announced its intention to deploy more troops in its northern regions. It has also signed an agreement that will allow U.S. troops to train in Norway's Arctic.

Russia says it will establish a permanent Arctic force. Russian general Vladimir Shamanov has announced paratroop drops on the North Pole. Just "symbolic," he claims: "We do not intend to engage in [saber] rattling." The message, though, is clear and is reported across the web, accompanied by ads for "RussianLoveMatch.com. Join Free."

Denmark has announced plans to create an Arctic command with a combined-forces military contingent for Greenland. "We're not going down a road towards confrontation," Brigadier General Joergen Jacobsen said. "Indeed, we're going down a road towards cooperation and collaboration."

The road it is actually going down is maximum exploitation of Greenland's offshore oil resources, which in August 2010 brought the Danish navy into Baffin Bay along the west coast of Greenland to protect oil exploration rigs. As the Danes presided over the debacle that was the Copenhagen climate change negotiations, they were completing plans to drill four oil and gas exploration wells off Greenland's Disko Island, smack in the middle of iceberg alley. Eight months later, a British company called Cairn Energy PLC claimed to have struck oil (or what it cautiously called a "working hydrocarbon system") with its first well. What the company found was oil-bearing sand, similar to Canada's tar sands, which it hopes is proof of the presence of liquid crude in the area. The find immediately brought the confrontation General Jacobsen denied would occur. A Danish frigate stopped the Greenpeace vessel *Esperanza* from entering a 500-meter exclusion zone it had set up around the drilling rig. Mindful of the BP deep-water blowout disaster in the Gulf of Mexico, protesters picketed Cairn's Edinburgh headquarters and the Bank of Scotland, which bankrolls the drilling.

More serious opposition could come from Canada. Two of Cairn's leases border Canadian territorial waters off Baffin Island. A third lease farther south also borders Canadian waters. Cairn Energy shares this lease with the Canadian oil and gas company Encana. The Greenlandic state oil company, Nunaoil, keeps a minority stake set at either 8 or 12.5 percent of all Greenland oil leases. The Cairn leases alone cover an area of about 72,000 square kilometers. Nunaoil has a total of twenty-three leases off the west coast of Greenland, from the Talbot Trough along Ellesmere Island down to the southern tip of Greenland—an area bigger than the North Sea. Ten of its leases border Canadian waters. If oil is dis-covered in any bankable amounts, confrontation between Denmark and Canada is almost inevitable. Oil reserves do not respect borders. Furthermore, the Greenland exploration wells put added pressure on Canada to enter what would be a classic race for the black gold. If Canada's tar sands are any indication, the runners inevitably will trample environmental concerns.

There's certainly supposed to be a lot of oil riches in the area. The United States Geological Survey estimates the oil and gas reserves of offshore Greenland at 48.3 billion barrels of oil equiva-lent (oil and gas). About 17 billion of these barrels, if they exist, are in territorial waters shared by Canada and Denmark.

If Greenland strikes oil, it could mean curtains for Denmark as an Arctic power. As a U.S. embassy cable from Copenhagen stated on November 7, 2007, "Greenland is on a clear track toward independence, which could come more quickly than most outside the Kingdom of Denmark realize." The cable went on to state, "One senior Greenlandic official commented recently that his country (Greenlanders and many Danes alike routinely refer to Greenland as a "country") is 'just one big oil strike away' from economic and political independence." The cable speculates that an independent Greenland would offer a "unique opportunity" for the United States to "shape the circumstances in which an independent nation may emerge." The cable notes that the U.S. has "real security and growing economic interests in Greenland . . . American commercial investments, our continuing strategic military presence, and new high-level scientific and political interest in Greenland argue for establishing a small and seasonal American Presence Post in Greenland's capital as soon as practicable."[4]

The Americans are making the most extensive review of military operations of any of the Arctic Five. Admiral Jonathan W. Greenert, U.S. vice chief of naval operations, on May 15, 2009, announced the establishment of "Task Force Climate Change" to develop an "Arctic roadmap for the Navy and then later for global climate change responses more generally." The 35-page roadmap was published in October of that year and provides the first look at how the United States Navy will develop new strategies and missions and assess weapons needs for an Arctic fleet as a result of changes in the Arctic environment brought on by global warming. Specifically, it will address "undersea warfare, expeditionary warfare, strike warfare, strategic sealift and regional security cooperation."

Canada, on the other hand, has talked tough but carried a twig. Since 2006 it has announced a list of measures designed to assert sovereignty over its Arctic archipelago. These include plans to build a "world class Arctic research station," a new military training center at Resolute Bay, a navy base on Baffin Island, and new Arctic frigates and icebreakers. But by the summer of 2010, the nation with the largest Arctic coastline had not taken any steps to execute these plans. No ships had been commissioned, never mind built, and the only permanent military force up there was not really a military force at all but a shadow group of Inuit militia called the Rangers.

I first traveled with the Rangers in 1982 near Resolute Bay, where a British officer and a Canadian officer were putting them through their paces. At the time, Canada had nothing else up there in the way of defensive capability other than radar stations. About fifty Rangers had gathered for some late summer training in temperatures just above zero. Uniforms were in short supply. That is, they didn't have any. A few had baseball caps with a Ranger insignia. The troops were all Inuit men who were paid a small fee to participate. They had to supply their own snowmobiles or all-terrain vehicles, if they had any. Most didn't. Their weapon was and still is a .303 bolt-action Lee-Enfield rifle that dates back to the Second World War, the kind of rifles you used to be able to pick up for fifty bucks out of a barrel at the local Army Surplus Store. The Rangers like them because they're durable and reliable in the cold. They don't jam up at minus-30.

The Department of National Defence gave each Ranger two hundred rounds a year, which the Inuit said they used for hunting. Their mission was to spot intruders and act as scouts if the Canadian army came up to take out some Soviet invaders or anybody else who tried to dispute Canadian sovereignty. Nobody took it seriously. Least of all Canadians. The Rangers were simply tools to patrol Canada's paper sovereignty. Because they were up there, Canadians could claim they used the territory and patrolled it and therefore it was theirs.

The Canadian officer, a captain, looked shabby in his ill-fitting uniform and spent most of the time chain-smoking. He wore his Arctic assignment like a punishment. The British officer, on the other hand, was well turned out and parade ready. He said he was on an exchange program, getting some Arctic exposure, but my guess is he was there to size up Canadian efforts at establishing sovereignty in the Queen's Arctic domain. Both officers spent their time displaying the outlines of Russian and American submarines on large white pieces of plasterboard so the Inuit could learn to distinguish class and country if they ever saw one surface through the ice. A few Inuit told me they had seen subs in the area. One of them said he thought he had spotted a Russian sub, but he wasn't sure.

Canadian subs didn't figure into the equation. Canada still doesn't have any that could go to the Arctic, at least not under the ice for long periods of time. That's why, when they do venture north, they stick to wide channels and leave well before the winter ice sets in.

Almost thirty years later, Canada has expanded the number of Rangers to 4,250 in 169 patrols across the Arctic. The goal is to reach 5,000 by 2012, which would include a sizable proportion of the 25,000 eligible Inuit living in the North. Their uniform now includes a red sweatshirt with the green Ranger insignia—a .303 rifle and ax crossed over a sprig of three red maple leaves. The Ranger mission statement: "Provide lightly equipped, self sufficient, mobile forces in support of the Canadian Forces' sovereignty and domestic operation tasks in Canada." Their motto: The Watchers.

The weakest link in Canada's Arctic claims is our presence patrolling Arctic waters, including the Northwest Passage. Since the 1970s, Canada has debated whether to build a Polar class icebreaker that is capable of year-round operations in the Arctic. Such cutters can ram through ice up to six meters thick, which is at the high end of winter sea ice thickness in the Arctic. Canada's largest breaker is the *Louis St. Laurent*, which holds the naval classification "heavy ice breaker." It is capable of breaking ice only about one meter thick and operates in the Arctic only in late spring, summer and fall. In 1985, the Canadian government budgeted $700 million for construction of a Polar class icebreaker for year-round Arctic operations, but it canceled the program five years later for lack of money. It was revived in 2008 with the promise to spend $720 million on a smaller Polar class breaker, to be christened the CCGS *John G. Diefenbaker*. But as of this writing, no contracts have been awarded and the delivery date has been postponed to 2017. Money continues to be an issue, which is strange considering the enormous wealth produced each year by Canada's resource companies. Net revenue from just one of those companies—Suncor Energy—over the last five years could have paid for fifteen new Polar class icebreakers. Most of its profits come from tar sands operations, which pay some of the lowest royalties in the world. In addition, Canada's annual subsidies to the fossil fuel industry total about $2 billion. So while companies such as Suncor gush profits using resources owned by Canadians, Canada can't seem to come up with the money to protect its own sovereignty.

So Canada uses less expensive methods, whose effectiveness is at best dubious. The government announced in June 2010 "an important measure to protect and defend Canada northern sovereignty." It amounts to the implementation of an honor system requiring that all domestic and foreign vessels of more than 300 metric tons report to the Canadian Coast Guard before entering Canada's Arctic waters. "This government, under the leadership of Prime Minister Stephen Harper, has taken unprecedented action to protect Canada's North, and today's announcement will allow the Canadian Coast Guard to keep closer watch on our Arctic waters," Gail Shea, the minister in charge of the Coast Guard, said. "With mandatory reporting, the Canadian Coast Guard will be able to promote the safe navigation of vessels, keep watch on vessels carrying pollutants, fuel oil and dangerous goods, and respond quickly in the event of an accident."

There are 36,563 islands covering 1.4 million square kilometers with a total coastline of 137,000 kilometers in the Canadian

Arctic archipelago. Canada has only two icebreakers capable of operating in the Arctic for only six months of the year. Shea could not explain how Canada plans to keep watch on these foreign or domestic vessels and quickly respond to accidents with such meager naval resources and little hope of new breakers anytime soon. For Canadians, it seems, the Arctic is little more than pride of ownership. It's snow and ice. It's a polar bear carved into a diamond. It's walruses and Inuit. A beautiful picture in a magazine. An inukshuk. For their governments, it's a place without a vote. It's a resource to be exploited. A vision to protect. A stewardship to feign. A vast conceit.

Other countries are not so slow. The United States is planning to replace its current fleet of polar icebreakers. Even the state of Alaska is considering using its oil revenues on two Polar class breakers. And then there is China.

China's aggressive Arctic policy calls for the construction of new icebreakers. For the last decade it has launched about twenty-six expeditions each year to conduct extensive research on ice in the Arctic—in addition to its projects in the Antarctic—using the *Snow Dragon*, the largest non-nuclear icebreaker in the world. At 21,000 metric tons, it has twice the displacement of Canada's largest icebreaker and is forty-three meters longer. It sports seven laboratories and carries three small-craft vessels, a helicopter and the latest in scientific and navigation equipment. In 2009, China decided to build a new, smaller polar research icebreaker to sister *Snow Dragon*. Delivery date: 2013.

Chinese scientists have been researching Arctic ice since 1995, when they hiked from Ellesmere Island to the North Pole. In 2004, China established a permanent research station in Ny-Alesund on the High Arctic archipelago of Svalbard.[5] Those islands are the sovereign territory of Norway and have become the base for China's Arctic research. They are also the location of the world's seed bank. Seeds from around the world are preserved in one of three permafrost vaults cut 120 meters inside a mountain near the town of Longyearbyen.

The islands' history provides a lesson for modern times. Northern European fishing and whaling nations had for years disputed their ownership, but nobody lived there, so no country could establish permanent status. Only when coal and other minerals were discovered was there a need to stake a claim and make rules. The Spitsbergen Treaty—also known as the Svalbard Treaty—of 1920 awarded sovereignty over the islands to Norway, but gave other signatory nations the right to live on the islands and establish commercial enterprises and research stations. So while Norway refs the game, anybody who signed the treaty can play. So far, only Russia and Norway have mined the islands, albeit under environmental conditions imposed by Norway. Ten countries have research units on Svalbard. China became one of about forty participatory nations when it signed the treaty in 1925; Chinese scientists thus have the right to be here and gather ice and weather data and, if they want, prospect for ore. They are headquartered in a large red two-story rectangular building guarded by two five-foot-tall alabaster Chinese lions.

China's signature on the Svalbard Treaty could also give it a claim to offshore resources, which pinpoints a key Arctic dispute. Norway, backed by Canada, claims the Svalbard continental shelf is part of mainland Norway and therefore not covered by the Svalbard Treaty. Russia and China disagree.

China's polar research is centered at the Polar Research Institute of China in Shanghai. *Snow Dragon* is the institute's ship. But many other Chinese institutes also perform extensive polar research. Their work was compiled in 2009 into a major report on Arctic issues that included topics such as Arctic law, politics, diplomacy, military issues, shipping, natural resources, Arctic societies and the environment. A main research interest is how Arctic melting will affect the Chinese coastal communities and low-lying regions. On August 7, 2010, *Snow Dragon* sailed through the Bering Strait and headed towards the North Pole. It fought its way through the Arctic ice and stopped at 86 degrees and 55 minutes north latitude, 178 degrees and 53 minutes west longitude, or about three hundred kilometers short of the Pole. Scientists planted the Chinese flag on the ice and set up a fixed ice station with a lime green domed metal module they call the "apple house," designed to protect the researchers against polar bears as they probed the ice. They spent fifteen days setting up a long-term unmanned observation station before heading home.

Commercial shipping opportunities also are of vital interest to the Chinese, who are charting two main shipping routes through the Arctic. One is through Canada's Northwest Passage, to bring Chinese goods to the North American eastern seaboard. The other is through Russia's Northeast Passage, which extends from the Bering Strait, along the coast of Siberia to Norway. This passage is already proving navigable. In the summer of 2010, Russian nuclear icebreakers led a small convoy of freighters through the passage without incident. At the present rate of ice melt, they may not even need icebreakers by 2020. An estimated 46 percent of China's economy is reliant on shipping. Both routes would substantially reduce shipping times. The Siberian route to Europe is shorter by about six thousand kilometers than the route through the Strait of Malacca and the Suez Canal. What's more, there are no pirates to increase insurance costs. At least not yet.[6]

Environmental groups such as Greenpeace and the World Wildlife Federation favor a global approach to Arctic governance. They want an overarching treaty for the Arctic and a stop to all mining and drilling until international rules of engagement can be established. The A5 says, Fat chance. These are our shores and our continental shelves, and we'll decide what happens up there. We've got the Arctic Ocean surrounded.

When I first heard about the Arctic Five meeting, it struck me as an example of a group of nations falling into what behavioral psychologists call a "social trap." This is where a group will pursue short-term gain that in the long term will make them all losers.

Here in the Gatineau Hills, the ministers were beginning the process of setting the rules for exploitation of the Arctic's many resources, but most of all its fossil fuel reserves. The burning of these reserves will inevitably lead to the continued contamination of the atmospheric space with greenhouse gases that ultimately could put at risk a good part of the earth's species, including us. That this is the logical conclusion of a substantial part of the world's scientific community is something that we cannot ignore. But here at this lush country estate, that is precisely what happened.

The logic of the ministers' thought processes is in one sense correct. Climate change—or "altering weather patterns," as the Canadian officials call it—is threatening to open up the Arctic to maritime traffic and resource exploitation. So it is important to lay down the rules that will best protect the interests of the individual nations. Given modern society's inclination to exploit available energy resources, the ministers are only acting in a compellingly prudent manner.

And therein lies the problem. "Prudence is a selfish virtue."

That's what English mathematician William Forster Lloyd noted when he discussed the problem of the "social trap" in two lectures entitled "On the Checks to Population" delivered in 1832 at Oxford University.

Lloyd was furthering a debate about the limits of population growth. His contribution was to argue that an individual will not take the future consequences of his actions into account if he knows that those consequences will be dispersed throughout the population. In other words, he will choose to seek an immediate benefit for himself even though he knows that his actions will come to haunt the rest of society. The pleasure of sex will trump any concern about the long-term effects of overpopulation on society as a whole. Society, of course, now has the reproductive technology to solve that problem if an individual chooses to use it. This is not the case with climate change, where, so far anyway, we have no technological solution other than abstinence from the burning of fossil fuels, and we don't seem to be keen on even a small amount of that. Society remains unflagging in its almost pathological pursuit of material self-interest.

"The future is struck out of the reckoning when the constitution of society is such as to diffuse the effects of individual acts throughout the community at large, instead of appropriating them to the individuals by whom they are respectively committed . . ." Lloyd wrote. "Where the consequences are to fall on the public, the prudent man determines his conduct by the comparison of the present pleasure with his share of the future ill."

Lloyd's lectures fell into obscurity until Garrett Hardin, a microbiologist from Texas, resurrected the discussion in a 1968 article in *Science* where he coined the term "the tragedy of the commons." He took his definition of "tragedy" from the philosopher Alfred North Whitehead: "The essence of dramatic tragedy is not unhappiness. It resides in the solemnity of the remorseless working of things." In present times this could be the remorseless workings of a society whose members share only the desire to pursue individual material wealth, what Whitehead called "the inevitableness of destiny."

Here's Hardin's tragedy:

The tragedy of the commons develops in this way. Picture a pasture open to all. It is to be expected that each herdsman will try to keep as many cattle as possible on the commons. Such an arrangement may work reasonably satisfactorily for centuries because tribal wars, poaching, and disease keep the numbers of both man and beast well below the carrying capacity of the land. Finally, however, comes the day of reckoning, that is, the day when the long-desired goal of social stability becomes a reality. At this point, the inherent logic of the commons remorselessly generates tragedy.

As a rational being, each herdsman seeks to maximize his gain. Explicitly or implicitly, more or less consciously, he asks, "What is the utility *to me* of adding one more animal to my herd?" . . . The rational herdsman concludes that the only sensible course for him to pursue is to add another animal to his herd. And another . . . But this is the conclusion reached by each and every rational herdsman sharing a commons. Therein is the tragedy. Each man is locked into a system that compels him to increase his herd without limit—in a world that is limited. Ruin is the destination toward which all men rush, each pursuing his own best interest in a society that believes in the freedom of the commons. Freedom in a commons brings ruin to all.

Whether the world is finite or not has no impact on the risk of catastrophic climate change except if we have the option of escaping into space. And, as Hardin reminds us, that is a fiction.

The Arctic Five claimed they have no option but to prepare the rules of exploitation in the Arctic. Given the undeniable progress of climate change, as admitted by all five countries, it is only wise that they do their best to construct rules aimed at avoiding catastrophes that could pollute their Arctic lands. In this sense, the imposition by this group of rules for development and navigation is prudent. A deep-sea drilling platform that springs a leak like BP's Gulf of Mexico disaster will harm Canadian or American or Russian or Norwegian or Danish (Greenland) coastal waters but not those of, say, the United Kingdom, where BP is headquartered and to where its profits flow. So it's only normal that these countries prepare for such events.

But that is assuming they have only one option on the table: development. Which is not true. There is a second option: leave the Arctic fossil fuels in the ground. Pledge, in other words, not to exploit them. That would mean that these five countries pledge to chart a new course for the world that steers away from further polluting of the atmospheric space, a space we once thought was inexhaustible but we now know is not.

This is not an option that any of these nations has taken seriously or discussed. This is because they are so consumed by Whitehead's "remorseless working of things"—in this case, the cosmic wheel of materialistic self-interest, personal wealth accumulation and economic competition. They feel that these are forces they cannot stop even if they wanted to. To paraphrase Hardin, each country is locked into a system that compels it to increase its wealth without limit.

Lloyd would probably say this is a classic case of the social trap. Most of the immediate benefits of the exploitation will accrue to the five players while the negative impacts will spread right around the world and will be realized only at some as yet unknown time.

If the Arctic Five are to run the show in the north, they first have to resolve border disputes and territorial rights over their continental shelves. This latter issue promises to take the international Law of the Sea to a whole new level.

Most pressing has been the dispute between Norway and Russia. For forty years the two countries had tussled over ownership of the potentially oil-rich continental shelf covering 176,000 square kilometers under the Barents Sea. The ice melt has brought this dispute to a head.

Both countries are eager to tap into these gas and oil fields. Russia is already constructing four 70-megawatt nuclear power plants on barges, each containing two reactors, to power oil and gas rigs in the Barents and Kara seas as well as service an estimated thirty-three remote communities along the full breadth of the Arctic coastline. The first was completed in 2011. Russia plans to build as many as fifteen floating nuclear reactors a year.

Since the 1970s, Russia has developed an expertise in year-round shipping in the Arctic. In 1978 it created an Arctic transport system to ship nickel from the port of Dudinka, in north-central Siberia, 231 nautical miles to Murmansk year-round. Nuclear ice-breakers carved the way through the ice. Later, with help from the Finns, it employed ore-carriers with icebreaking capacity and no longer needed the icebreakers. It also employed shallow-draft nuclear icebreakers for year-round Siberian river navigation. (In 1978, Canada built the MV *Arctic*, an icebreaker used to carry ore from zinc and lead mines on the High Arctic island of Little Cornwallis, west of Resolute, and in Nanisivik on the north end of Baffin Island.)

Russia's nuclear legacy has already contaminated parts of the western Russian Arctic. Spent fuel and radioactive liquid and solid waste from 198 obsolete nuclear submarines, mothballed nuclear icebreakers and thermoelectric generators are poorly stored, primarily in three thousand containers in Andreeva Bay, close to the Norwegian border, as well as more down the coast in Gremikha. These sites are "severely contaminated," according to a 2009 study by the Arctic Council. In 1982, radioactive water leaked out of the containment area and into the sea.[7]

As Western countries are paying to help clean up these derelict warehousing sites, Russia's new floating nuclear plants have become a major worry to Arctic nations. The plants "raise issues about how waste will be handled and about increased marine transport of spent fuel in the Arctic," the Arctic Council says. "These power plants would represent new potential sources and may increase risks of radioactive contamination."

The Soviet Union dumped nuclear waste from its bomb test sites in the Kara Sea. Monitoring of these undersea waste sites has been sporadic at best and the Arctic Council worries that climate change could help release radioactivity: "The impact of weather and climate on infrastructure is well known for Andreeva Bay, where freeze-thaw actions contributed to loss of integrity of the fuel storage facility and extensive contamination of the Andreeva Bay site. Further degradation combined with precipitation has contributed to radioactive material being washed out into the marine environment."

The nuclear threat, however, is not a high priority when measured against the urge to develop Arctic oil and gas reserves. Soon after the meeting in Gatineau, Russia and Norway reached a settlement, agreeing to divide the disputed Barents Sea area equally. They also agreed to develop jointly its oil and gas fields. Various surveys estimate the area contains between 7 and 29 billion barrels of oil plus trillions of cubic feet of gas. On September 15, 2010, Norwegian prime minister Jens Stoltenberg and Russian president Dmitry Medvedev both traveled to the Arctic city of Murmansk to sign the treaty settling the forty-year-old dispute.

Both sides had already agreed to joint ventures in undisputed territory. Their national oil companies have negotiated joint plans to develop Russia's Shtokman gas field, whose estimated reserves of 141 trillion cubic feet make it one of the largest fields in the world. Shtokman lies about 430 kilometers offshore from the Norwegian-Russian border in water that is about three hundred meters deep, and threatened by icebergs and vagrant sea ice. These dangers have delayed exploitation. Russia needs the offshore technology Norway has perfected to exploit the wells. The deal with Statoil, Norway's state-owned oil company, gave Statoil a 24-percent stake in the field. France's Total S.A. has 25 percent while Russia's Gazprom holds 51 percent.

It is a US$20-billion project that Norway needs. Its North Sea oil and gas wells are swiftly depleting. It's relying heavily on development of Arctic energy reserves to continue the flow of cash that has made Norway one of the richest countries in the world, with a petroleum pension fund that totaled US$443 billion as of March 2010. The fund, the fourth-largest such fund in the world, is the largest holder of corporate shares in Europe.

Norway's partnership in the Shtokman gas field venture comes with the risks attached to allying itself too closely with Russia, and the deal reflects the already strained and distrustful atmosphere that pervades the exploitation of this new world. So far, however, the lure of Arctic gas trumps the risk, which has less to do with the potential of the gas field than with Russian politics and corruption. Events leading up to the deal with Norway indicate that Russia is once again using its energy resources as a political pawn in the game of international trade.

Russia initially had sought partnerships with American companies, including ExxonMobil and ConocoPhillips. At the time, Russia was trying to persuade the United States to back its entry into the World Trade Organization. Membership, it believed, would lead to greater trade and a more diversified economy. Russia relies heavily on oil and gas, which already enjoys global free trade status. When the Americans balked, Russia turned back to the Europeans. But the game was not over. The Arctic represents the last exploration frontier and Western energy companies are desperate for an advantage, particularly since they have been locked out of so many countries that give priority to their national oil companies. When Russia turned to Statoil, it delivered a lesson to the U.S.A. in modern oil politics. A year later, with the promise of greater U.S.

participation in Russian energy development in mind, President Barack Obama decided to back Russia's WTO bid.

One question that concerns Norwegians is what will happen after Statoil makes its initial investments and technology transfers to get the wells flowing. Russia will then hold all the cards. With the wells in place and operating, Norway's fear is that Russia will want to renegotiate the contract to Gazprom's advantage or even push Statoil out of the deal, as it has done with Shell Oil and other Western companies in the Sakhalin field, on Russia's east coast.

Norwegian companies have certainly felt the pain of doing business in Russia. Small Norwegian oil firms have been driven to the edge of bankruptcy because of their Russian investments. Norway's state-controlled telecommunications company, Telenor, has since 2005 been enmeshed in a string of billion-dollar law-suits launched in Russia and Ukraine by its Russian partner Mikhail Friedman, chairman of Alfa Group.[8] One lawsuit ended up in a New York courtroom, where the judge called it "vexatious" and a "sham." Transparency International rates Russia as one of the world's most corrupt countries in which to do business, a place where government, courts and businessmen have a reputation for colluding to fleece foreigners when they are not defrauding each other.

Statoil is an independent company but is ultimately controlled by the Norwegian government. This brings a strong element of political risk to the deal. Norway's reliance on energy revenues from Russian wells could limit Norway's foreign policy choices by linking the country too closely with Russia. "A situation may arise where the Norwegian state feels subject to pressure to make political concessions to support the commercial interests of Statoil," Norwegian political economist Morten Anker has written. "A Norwegian company's involvement in a Russian mega-project may have strings attached. An even more subtle result could be that the Norwegian government, conscious or not, shifts its policy towards Russia in order to avoid a pressure where Statoil's position in Shtokman could be at stake—as a preventive action."[9]

There's no doubt that Norway harbors deep distrust of Russia, but its depleting oil wells in the North Sea have forced an alliance. With more than half of its exports coming from oil and gas, Norway's economy is chained to fossil fuels. It needs the large revenue stream promised by the partnership with Russia. Still, Norway remains cautious. It announced its intention to redeploy troops into its Arctic. In a classic understatement, Norway's foreign minister, Jonas Gahr Store, told the media just before the Gatineau meeting that maintaining "relations with Moscow is a complicated matter, because Russia is not quite normal." He added: "It is not yet a stable, reliable, predictable state." When I asked him why, in that case, Norway is ready to make a multi-billion-dollar investment with such a risky partner, he got a little defensive and backtracked. "Norway is Russia's neighbor. We have a lot of very constructive cooperation on Arctic issues, on border issues, on Barents Sea issues, and the large majority of that story is the story of close cooperation."

Oil and gas is why the struggle for sovereignty in the Arctic has centered on continental shelves. They are important because, as is true of continental shelves almost everywhere in the world, they are underlain by potentially petroleum-bearing sedimentary rocks. The broad continental shelf that extends off the northern coasts of Russia and Norway has already proved to be an oil- and gas-rich area. The more constricted shelf in the Beaufort Sea off Alaska and Canada has even greater potential, primarily in the shallow waters off the Mackenzie River Delta. The rest of the Arctic is still a guessing game.

The challenge for these five countries is to claim the undersea regions that extend well past the standard 200-nautical-mile limit. All five countries are making vast claims. Should they all be granted, they will give them jurisdiction over more than 88 percent of the

Arctic waters.[10] Which is why they are working closely together towards that goal even while competing with each other over conflicting claims.

Under the United Nations Convention on the Law of the Sea (UNCLOS), a country can claim up to 200 nautical miles offshore from the shoreline at low tide. However, if the continental shelf extends farther, a country can claim up to an additional 150 nautical miles. And under certain geophysical conditions, a country can claim well beyond that. This is where the law has been somewhat cryptic and where the Arctic Five are pushing the legal boundaries for all they are worth.

How to define this undersea geography is the question. Part of the problem lies in the definition of a continental shelf itself. The Law of the Sea is based on a definition from 1975. Undersea mapping technology has improved considerably since then, complicating the situation. The famous Lomonosov Ridge, for example, which runs across the North Pole from the Russian continental shelf to the Canadian-Greenland shelf like a half-buried dinosaur skeleton, can't be a ridge if the Russians want to use it to extend their claim to the North Pole. To grab the Pole position, they have to prove it's a shelf.

Continental shelves are sedimentary extensions of continents. They are made of sediment dragged off the land by the angry oceans or dumped into the sea by river systems. But in the main they are sediment pushed towards the sea by glaciers during the last ice age. As the glaciers melted and receded, this land disappeared under the rising seas. Essentially, the Arctic Five are trying to claim that land back. The basic principle is that what comes from the continent belongs to the continent. What comes from the sea belongs to the sea. What belongs to the sea is "common heritage." Advances in undersea mapping have allowed scientists to show that a great deal more of the seafloor than previously thought consists of sediment from the continents, which defies a widespread and convenient belief on the part of non-Arctic countries that the Arctic's resources are common heritage. It's a potential mess of claims and counterclaims.

A 2009 decision by the Commission on the Limits of the Continental Shelf on a Norwegian claim for territorial rights— over large sections of the Norwegian Sea between Norway and Greenland and a section north of the Svalbard Arctic islands— broke the game wide open. Previous mapping had shown these areas to be totally without continental sediment. In this ruling, the commission recognized Norway's claims to sections of the continental shelf where the sediment had fanned out thousands of square miles over the seafloor.

The biggest prize is the Lomonosov Ridge. Initial exploration data indicates it could be an undersea treasure. Coring samples and seismic data taken in the summer of 2004 "suggests that most of the key elements for petroleum may exist on the Lomonosov Ridge."[11]

Dispute over the ridge targets the issue of who owns the North Pole. Russia's claim would give it control of the High Arctic Ocean and 120 million hectares of seabed in a wedge from Russia's eastern and western boundaries to the North Pole. It amounts to more than one-third of the entire Arctic Ocean pie. Russia claims the Lomonosov is a natural extension of its continental shelf and not a seafloor uplift. Greenland and Canada agree that it's not a ridge, but they claim it as part of their shelf.

The Law of the Sea says the ridge has to be a "natural" extension of a continental shelf to give any of these countries the sole right to exploit its riches. In 2002, the Commission on the Limits of the Continental Shelf rejected the Russian bid for ownership of the ridge. But in 2007, a privately funded expedition of Russian Arctic scientists, along with rich Russian, Saudi and Aussie adventurers who paid for the privilege, sailed a nuclear icebreaker to the North Pole, cut open a two-meter-thick section of ice and dropped two manned submersibles through the hole. They sank down about three kilometers to the ocean bottom, where they picked up sedimentary samples. Before returning to the surface, they left behind a small titanium Russian flag and declared the North Pole Russian. The Russians had achieved something nobody else had done—cut through the permanent polar ice shield, dropped down to the sea bottom, planted their flag and returned to tell the tale.

The United States called the flag planting "provocation." Canada's then foreign minister, Peter MacKay, boyishly complained: "This isn't the fifteenth century. You can't go around the world and just plant flags." Russia's then president Vladimir Putin told everybody to calm down: "Don't worry. Everything will be all right. I was surprised by a somewhat nervous reaction from our Canadian colleagues. Americans, at one time, planted a flag on the moon. So what? Why didn't you worry so much? The moon did not pass in the United States' ownership." But that same month, Russia announced the decision to resume long-range bomber runs over the Arctic.

The rock samples and ridge mapping from this polar venture produced fresh evidence of sedimentary rock similar to Russia's continental shelf. The scientists claim that this proves the Lomonosov Ridge is Russian. The UN is currently dealing with this new claim.

Canada has three disputed boundaries. Two are with Denmark. One is over ownership of an ice-bound rock called Hans Island in the Nares Strait between Greenland and Ellesmere Island. Geologists regard this dispute as a mystery since "nobody imagines that there are any oil and gas resources to be found anywhere near this tiny island—the geology is all wrong," according to Andrew Miall, a petroleum geologist at the University of Toronto. The other dispute with Denmark is over 200 square kilometers of continental shelf under the Lincoln Sea off the tips of Ellesmere Island and Greenland, which even the overly optimistic United

States Geological Survey estimates has little oil or gas potential. Canada's third dispute is over about 22,000 square kilometers of potentially oil-rich ocean floor under the Beaufort Sea off the coast of Alaska and the Yukon. Canada wants to extend the Alaska–Yukon border along the west 141st meridian 370 kilometers due north towards the Pole. The United States wants the line to run the same distance but more perpendicular to the Canadian shoreline, thus giving the Americans a bigger chunk.

The settlement of the dispute will largely depend on the work of a new torpedo-shaped yellow submersible called *Discovery*. The unmanned Canadian marine drone is programmed to map and measure with multi-beam sensors (radar, sonar) the nature and thickness of the sediment. The spring of 2010 was its first real test.

I met Stephen Nichio, one of the engineers who designed *Discovery*, in the Polar Shelf hangar in Resolute. He was in the process of packing pieces of the drone into thirty-two wooden crates for the trip south to Port Coquitlam, British Columbia. He had camped for about a month on pack ice on the south side of Borden Island, halfway up the west side of the Canadian archipelago. From there he and his team had lowered their sub through the ice and released it on a voyage that took it about 322 kilometers around the island and out over the continental shelf to where members of the Geological Survey of Canada waited for it to nose up through a hole in the sea ice near their floating camp.

"The guys were up in the remote camp and they were waiting and waiting and waiting," Nichio said. "Then they got a return [message] from the acoustic telemetry on the sub and it says, 'Beep, beep, I'm here.' They were pretty delighted."

It was about forty-nine meters off target, but when you think that the ice was traveling at a rate of about twenty kilometers a day, it was pretty impressive that *Discovery* still found the hole.

"We have designed an underwater charging system [so] that we can capture the vehicle and once we plug it in . . . we have Ethernet telemetry and we can charge and reprogram the vehicle without taking it out of the water," Nichio said.

The drone is twenty-two feet long and slim enough to allow a really big man to wrap his arms around it. It cost about $1 million. On its side is printed: "If found please call 902–426–3100."

I asked Nichio if it has a habit of getting lost. He said it's programmed to work its way around the Arctic islands to arrive at the precise destination. "We sent it out on missions that took three days," he said. "We give it a mission from its starting point to the end point. That's like walking across Toronto with a bag over your head and reaching out at the end and clasping your door handle. But if something goes wrong and just in case somebody finds it, they know where to call."

Discovery's measurements will help define the extent of the continental shelf primarily around the western coast of the Canadian archipelago all the way up to the Lomonosov Ridge. Canada must present its findings to the UNCLOS and the Commission on the Limits of the Continental Shelf by 2013.

So, if all goes as planned, in about five years the Arctic Five could have locked up almost 90 percent of the Arctic Ocean and be able to pretty well call the shots over a land and sea area that takes up about 6 percent of the earth's surface. This would mean total control over resources and the right to regulate shipping in the continental shelf zones. Asian exporters would be at a distinct disadvantage. Nordic countries would gain new power and influence on the world stage. And the warming Arctic would feel the cold hand of Western corporate culture.

CHAPTER 7

THE ARCTIC'S POISONOUS SECRET

IN WHICH WE EXPLORE THE MYSTERIES OF ISOLATION AND TOUCH THE EMERGING THREAT EMBEDDED IN AN ICE CRYSTAL

THREE MONTHS AFTER VENTURING OVER THE FREEZING ICE CAPS of Devon and Ellesmere islands, I return to the High Arctic.

It is August 13, 2010. Temperatures are already beginning to cool. We are descending the back slope of a High Arctic summer. Three days ago I got a call from Captain Jennifer Jones at the Trenton air force base in Ontario telling me there was a seat available on a flight up to the base in Alert if I wanted it. I packed my gear and after a four-hour drive from Montréal I arrived in Trenton at five in the morning to catch a flight that would take me 4,380 kilometers north to Alert, on the northern tip of Ellesmere Island, where the Canadian military has maintained a small station since 1956 and where Canada has had a weather and atmospheric science station since 1950.

We fly first to Iqaluit on south Baffin Island, then to Resolute and finally on to Alert. On my trip in May, a rich mantle of snow and ice covered every square inch of this solemn place, but now the waters are blue and ice free and the land is brown and barren.

Only the mountain glaciers and the ice caps and ice fields glow a brilliant white, and there are not too many of those. And only when we near Alert do I see the packed Arctic sea ice that guarantees its isolation.

Accompanying me are two air force explosives specialists coming to Alert to inspect the safety of the station's five magazines. There are also several civilians, including a food supplier, a plumber and an electrician come to repair the septic system and water purification plant. (As I will discover, there is no septic system to speak of. A pipe dumps the sewage directly into the ocean. The government is just installing a better pipe.)

Alert is a high-security military listening and communications post where a handful of scientists also gather weather and climate change data. It is considered the world's most northerly community, yet it is home to no one except the sixty to sixty-five soldiers and civilians who are simply passing through and maintaining the base in a largely symbolic effort to guard Canada's High Arctic sovereignty.

There is nothing symbolic, however, about Alert's elaborate listening devices and communications stations. The shortest path for long-range bombers and intercontinental ballistic missiles between Russia and the United States lies across the Arctic. Most of the Soviet Union's nuclear warheads were deployed in the north. During the Cold War, Alert provided assurance that any Soviet missile launch or bomber attack would be detected, with the promise of an immediate response. Alert's frontline position was critical to the Cold War's fundamental strategy of deterrence embodied in the threat of mutual assured destruction (MAD). "As long as both sides had confidence that an attack could be detected in time to launch their own missiles, the logic of mutual assured destruction was said to have maintained the peace," Rob Huebert, of the Centre for Military and Strategic Studies at the University of Calgary, says. Alert supplied that confidence. While its manpower is now reduced and satellites are the new detection .devices, its mandate hasn't really changed. It's still the Western democracies' listener in the High Arctic.

Alert supports a small number of technicians who maintain the computers and long-wave listening antennas. The data they accumulate is sent from a high-security, locked-down building (even the commanding officer doesn't have security clearance to enter) attached to the barracks, via five relay stations to Eureka, and then down to the military's intelligence and communications branch in Ottawa. The man who runs the operation is an expert in electronic warfare. During his tours in Afghanistan, his job was to disrupt Taliban signals, particularly those designed to trigger roadside bombs.

Alert's motto is *Inuit Nunagata Ungata*, meaning "Beyond the Inuit Land." Soldiers come here on six-month tours with a few weeks' furlough in the middle to break the grinding monotony of being stuck on an isolated high-security base that is accessible only by air. The nearest community is 676 kilometers due south, in Thule, Greenland. The nearest medical aid is fourteen hours away by plane, given that a plane has to fly up from Trenton and then return. The winters are pitch-black and minus-40 degrees Celsius and the region is covered in snow and ice. No one is allowed beyond the perimeters alone and without a radio. Blizzards can be as blinding as a desert sandstorm. When they hit, no one is allowed outside the barracks.

Safety lines link the buildings should anybody be caught in a storm, which can sweep through without warning. The summers are 24-hour daylight, with temperatures that hover on each side of zero, although more recently they have risen as high as 20 degrees Celsius. The base is littered with the accumulated junk of sixty years, including rusting barrels, antenna towers, fuel tanks, old trucks and pallets of used batteries. The only color comes from the hardy but delicate Arctic flora that includes the rusty purple, green and yellow hues of the lichen, the tiny lemon-colored Arctic pop-pies and a unique Arctic tree whose branches cling to the rocky soil.

Alert's buildings are giant sheet metal meat lockers in reverse: their refrigeration doors are designed to keep the cold out and the heat in. It costs about $80 million a year to run the base, most of which is spent on fuel for heating, electricity and the cost of flying in supplies. CC-177 Globemasters ferry in most of the cargo. Their four jet engines burn 20,000 pounds of fuel an hour, which works out to 280,000 pounds to fly from Trenton to Alert and back. The annual supplies for Alert are shipped each August from Montréal by cargo ship to the American base in Thule and then airlifted to Alert in an operation code-named Boxtop. Alert stocks enough dry food to last a year and, weather permitting, the air force tries to fly a plane in once a week with fresh produce.

We arrive in frosty polar weather of minus-2 degrees. It is misty and overcast, and the fractured pack ice is stacked up over Dumbell Bay and out into the Arctic Ocean as far as the eye can see. We land on the gravel runway, where base commander Major Brent Hoddinott and his small staff greet us. Not a lot happens on this base, so when we arrive at the main entrance hall of the barracks, which they call the "beach," another crowd shows up to applaud the new arrivals and bid farewell to those either ending their tour or heading out on leave, a triumphant moment given the extreme isolation of this tiny station.

Hoddinott is young, trim, over six feet tall and extremely fit, as I will find out later when we trek west over the hills towards the mountains. He's a chemical engineer, trained navigator and career air force officer with thirty tours in Afghanistan. Before Alert he was deputy commanding officer of 426th Squad. This is his first com-mand, which is why he likes it up here. "I get to be the big chief."

The air force runs the base, but it has no aircraft attached to it. Neither does it have any defensive or offensive capability. Its explosives are used in the rock quarry. Next week, Hoddinott says, there will be for the first time a flyover by two CF-18 Hornets as part of the annual joint Canadian-American-Danish Arctic maneuvers called Nanook. The F-18 has for decades been Canada's interceptor of the Russian bombers that routinely probe the edges of Canadian Arctic airspace. "There's not a lot of traffic up here, so it's something," Hoddinott says with a slight smile. The flyover is part of Prime Minister Harper's annual visit to the North. F-18s can't land because the gravel runway would shred their tires and Canada isn't prepared to shell out for a concrete runway. Hoddinott admits that the only real enemy, for the moment anyway, is the weather and cabin fever.

The High Arctic is an enormous place with nowhere to go. It can quickly become a prison. The isolation poses a challenge that plays with your head. Mentally you are resigned to your allotted tour. You work, you eat, you sleep, you exercise—maybe. The only Inuit I meet at the station is twenty-year-old Paul Nungaq from Grise Fiord, which is 725 kilometers south but in Arctic terms is the next village. Even he feels the isolation. He's one of four Inuit civilian apprentice mechanics who work three months in and then five weeks out. Like the others, he maintains a rigid pattern. Eats breakfast seven to eight. Works until noon. Eats from twelve to one. Back to work until five. Dinner from five to six. In the evening he drinks beer and plays pool. Then he's off to bed to start the cycle all over again. Others kill time watching television (horror films set on remote space stations are a big hit), reading, playing cards, working out in the gym or drinking in the mess. And when it's not minus-40 degrees Celsius and completely dark, you might go outside for a hike. But in the black darkness of an Arctic night that's out of the question. I'm told that most staff rarely venture much farther than the barracks, the cafeteria, the mess or the gym.

Shipping out creates its own anxiety. You suddenly are eager to leave, but your departure date is never certain. Delayed or postponed flights are common. So as your departure date nears, weather reports and rumors about goings-on at the airport take on extreme significance to the point of inducing panic. Soldiers scheduled to leave are often put on medical watch lists. The stress of the tour is not in the tour—it's in the leaving. Everybody knows the story about the desperate soldier who tried to walk home.

Ultimately, Alert has little to do with security in the Arctic. Its job is to protect the industrial and populated centers of the south. Its Arctic home is simply an outpost representing a southern concept of security. The comedy of sovereignty is played out by the Canadian prime minister, who in another part of the Arctic posed for the media cameras while standing on an ice floe with an army general at his side, then skipped stones across the open water. It is a mirage. Arctic security should be about protecting its environment, to which the Canadian government pays only lip service.

Seven kilometers from the base, at the end of a gravel road and far from possible contamination by the military station's exhaust fumes, sits a small mustard-colored metal building alone in the polar wilderness. Attached to one side is a three-story aluminum tower. Scattered around the building and decorating the tower is an array of electronic sensors and metering devices of all shapes and sizes that track solar radiation, greenhouse gases, the albedo effect, and persistent organic pollutants including mercury. Many universities and science institutions such as the National Oceanic and Atmospheric Agency have planted their equipment here. This station is the most northerly and among the most critical data collectors, taking the amplified pulse of climate change at its very heart, which is why I am here.

No vehicles are allowed near the building because their emissions would contaminate the scientific data. So Josh Benmergui, twenty-three, a student writing his master's thesis at the University of Waterloo, drives me to within a kilometer and we walk the rest of the way. Even in this remote area of the globe, the Canadian government has attempted to lock out the world from the reality of what is happening. Josh's boss, Chris Carson of Environment Canada, initially refused to allow any of her tiny staff to drive me out to the monitoring installation even though they go there every day. She finally relented when I told her I was going to walk. She agreed someone would take me there "but nobody will answer any of your questions." These were her instructions from Ottawa, she says. No information was to be released other than a photocopy of a two-page colored pamphlet containing government public relations contact numbers. Canada is a country where government-funded scientists agree to be muzzled. I told her I didn't come more than four thousand kilometers to look at a brochure.[1]

So my visit to the place is awkward and at times slightly comical. Inside the building are three small laboratories filled with the usual array of computers, wires, tubes and glass bottles. There's also a bedroom with bunk beds in case the weather strands any of the scientists. Benmergui sticks to the letter of the law. "I can't talk about anything science or policy related. I don't want to lose my job," he says rather sheepishly. I feel as if I have wandered into a temple of silence. Benmergui is both a victim and a tool of the government's clampdown on information leaking out about the severity of climate change in the Arctic. Like many Canadian scientists who rely on government and private industry for financing, he has been forced to make a deal with the devil. Our communications become so absurd that when I point to a machine labeled "CO_2 Analyzer" and ask him what it does, he refuses to respond. "I guess from the label it analyzes CO_2 content in the atmosphere," I suggest. He purses his lips, shakes his head and shrugs. We play this game throughout the visit. Finally, I ask him about his master's thesis. Major Hoddinott told me that Benmergui is working on the effects of climate change on mercury in the Arctic. But he won't discuss it either, though he does blurt out: "There's a lot of effects of climate change that you might not expect."

"Such as?"

"I can't say."

There is nothing secretive about what he won't talk about. The scientific community has been urgently probing these unexpected effects for a decade. The problem they confront is deadly, but the science is difficult and still unresolved.

Earlier in the day I had walked down to the shoreline at the end of the runway in search of one of those unexpected climate change effects. The temperature was minus-2 degrees Celsius. The azure sky was cloudless except for a distinctive white bank of fog working its way south along the horizon. Huge chunks of gleaming white sea ice shaped like giant mushrooms, tabletops or miniature mountains nudged up against each other in a sort of polite jostling. A hundred years ago this area was part of one giant ice shelf that had been there for at least five thousand years. The ice was dozens of meters thick and extended along the entire north shore of Ellesmere Island and out into the Arctic Ocean. By the beginning of the twenty-first century the shelf had split up into six separate ice shelves, and the one around Alert disintegrated completely. Two of the five have since broken up and the remaining three are not looking too sturdy. (In fact, on this very day, a 50-square-kilometer section of the Ward Hunt Island Shelf, about 170 kilometers west of Alert, was breaking up.[2]) The fractured pieces of ice pack I was about to climb onto are probably a combination of new sea ice and old ice broken off from the remaining shelves. The water around them was clear, cold and tranquil enough not to disrupt the wafer-thin sheets of fresh translucent ice floating there. Underneath this broken ice cover, according to the United States Geological Survey, there is between a 50 and

100 percent probability that we'll find 1.3 billion barrels of oil and 10 trillion cubic feet of gas equivalent.

But I hadn't come looking for petroleum. I had come in search of something that I was intending to discuss with Benmergui. I spotted a small ice floe close enough to the shoreline to step onto. This led me to another ice floe, where I found what I was looking for: the telltale filigree shapes of frost flowers—delicate, spiderlike crystals about an inch long that form on the upper edges of fresh sea ice. The thin, tangled lattice I found had grown more than a foot above the floe, reflecting the sunlight in all its brilliant colors. I extended a finger and lightly touched one of the crystals. It immediately collapsed, taking several others with it, falling almost feather-like into the sea. As I looked around at neighboring floes, I saw gardens of the crystal beauties.

They seem harmless. But their beauty hides a growing danger. Inside these crystals are concentrated all the elements of Arctic seawater, including massive amounts of bromine and mercury. A growing concern among scientists is that climate change has the potential to increase substantially the concentration of mercury—particularly highly toxic methyl mercury—in the oceans as well as in the freshwater lakes and wetlands of the Arctic and the boreal forest. The danger we face is nothing less than the poisoning of our food chain.

Methyl mercury is a powerful neurotoxin that enters the food chain primarily through fish and other marine animals. The amounts increase—bioamplify—as it progresses up the chain to humans. According to the United States Agency for Toxic Substances and Disease Registry, methyl mercury is rapidly absorbed into the bloodstream through the gastrointestinal system and then gets "trapped" in the brain. It is known to impair kidney function and neurological development, to damage cognitive thinking, memory, attention, language, fine motor and visuospatial skills and hearing. It also can lead to starvation. It accumulates in our bodies and is difficult to get rid of. The toxin is particularly dangerous to children. "Methyl mercury that is in the blood of a pregnant woman will easily move into the blood of the developing child and then into the child's brain and other tissues," the agency states.[3] This can adversely affect the development of a baby's brain and nervous system. The toxin is transferred to unborn children through the umbilical cord and can cause mental retardation. Unsafe levels of methyl mercury have been found in the umbilical cords of Inuit babies. Studies have found evidence of mercury poisoning affecting the nervous systems of Inuit children in northern Québec and in the Faroe Islands.

There is increasing scientific evidence that climate change is disrupting a natural and relatively benign chemical cycle; this disruption could unleash a major methyl mercury poisoning of our food chain. Scientists have only recently begun to explore the dangers that may lie ahead. The climate change station at Alert began tracking atmospheric mercury in 1992, which in climate change science is not very long ago. This research was part of a wider study of the content and effects of the ugly brownish haze that hangs over the High Arctic in spring, caused by pollutants migrating from the earth's industrialized centers. In 1996, a scientific report identified methyl mercury as an important threat to humans and urged scientists to investigate the natural processes that create it.

Mercury is naturally released into the environment from the breakdown of minerals in rocks and soil and from volcanic activity. A marked increase in environmental mercury has occurred over the last two centuries, primarily due to the burning of fossil fuels, mining, smelting and refining processes. Industrialization has increased by a factor of three the amount of mercury in the environment compared with preindustrial times.[4] Mercury levels in what we think of as the pristine Arctic are by far the highest in the world. Tests on Inuit whose diet includes fish and seal meat show they have higher levels of mercury than people living in industrialized cities. Polar levels reach ten to a hundred times those found in more southerly regions, which is counterintuitive as there is almost no industrial production in polar regions and very little human interference.

Ice cores show that most of the mercury is from the industrial era. The mercury found in the Arctic migrated there from southern industrial centers via atmospheric and ocean currents. But that does not fully explain the excessively high measurements of methyl mercury. Part of the answer to that mystery lies in a phenomenon unique to polar regions: "atmospheric mercury depletion events" (AMDE), first observed in 1995 in Alert by a Canadian government scientist named William Schroeder. These events are natural springtime occurrences that take place in all polar and even subpolar regions. "Wherever we look for it, we find it," Schroeder's colleague, chemist Jan Bottenheim, says. "It happens all over the Arctic."[5]

Prior to the discovery of AMDE, it was generally agreed that atmospheric mercury was a relatively benign form of gaseous mercury that remained in the atmosphere for a long time. In 1995, however, Schroeder noticed that the readings he was taking around Alert from March to June showed sudden and dramatic drops in atmospheric mercury. The graph lines plummeted and then almost as quickly hurtled back up again. Literally overnight, it seemed, the lower atmosphere was substantially depleted of its mercury content and then a few days later the level rose again. What was going on? Where did the mercury go and why did it suddenly return? Three years later, Schroeder published a paper detailing this phenomenon. The work unleashed a plethora of research.

What scientists have since discovered is alarming. In a mere 24-hour period or less, during the first polar sunrises and before temperatures climb above zero, explosions of bromine emitted by the saline brine on the surface of fresh ice oxidize the inorganic gaseous mercury in the atmosphere, transforming it into particulate or reactive mercury. This kind is heavier than air because it absorbs moisture, which causes it to fall immediately to Earth. The atmosphere in the Arctic is suddenly depleted of as much as half of its mercury. Much of it falls on the snow and sea ice that blankets the Arctic in early spring. Then ultraviolet light from the sun vaporizes an unspecified amount of the mercury back into the atmosphere. That its stay on the snowpack is short is a sort of saving grace for the food chain because it limits the amount of mercury that annually enters the ecosystem. The mercury that does not return to the atmosphere goes into the ocean and freshwater systems with the summer snowmelt, where it eventually is transformed into the deadly contaminant methyl mercury. Scientists are still not certain how much of the mercury remains on the snowpack to be ingested by the Arctic Ocean and wetlands, but several studies suggest that each year there is a substantial net gain of methyl mercury in the Arctic environment. The cycle continues each year as atmospheric and ocean currents transport industrial mercury emissions north. As a result, more and more methyl mercury accumulates in the food chain, which is a danger if only for the selfish reason that increasing amounts of our seafood come from the Arctic.

Researchers are still struggling to gain certain knowledge of exactly how much atmospheric mercury becomes methylated in the oceans and wetlands. They are also racing to find out how man-made climate change is making this phenomenon more powerful.

Alert is part of a global network of atmospheric monitoring stations and its data, including the data on atmospheric and land-based mercury, is distributed widely. Fortunately, the Canadian government cannot muzzle all scientists. So I spoke to Vince St. Louis, a professor at the University of Alberta who has spent his career studying mercury in the environment. More recently he has examined why some High Arctic and subarctic marine animals and freshwater fish contain concentrations of methyl mercury that are high enough to endanger the health of northern peoples.

Before climate change, he explains, the atmospheric mercury depletion event was not dangerous since the mercury that fell on the snowpack remained there for only four or five days before returning to the atmosphere. "Even though it falls out of the atmosphere, it really goes back to the atmosphere before it has a chance to get methylated," he says.

Warmer temperatures, however, mean more melting and more open water. Then the story takes a fresh twist and becomes still more disturbing. Work by St. Louis and others shows that when the mercury falls in open water, it sinks to lower depths, where scientists believe that bacteria from organisms such as plankton work on it and transform it into methyl mercury. The same is true if it falls in lakes or in stagnant wetlands rich in bacteria-producing plants.

"That process [of methylation] may not be instantaneous," St. Louis says. "That might be months or years down the road. It might take a while for the mercury that falls into an open part of the ocean to get transported down to deeper depths where it could get methylated." The melting of the Arctic sea ice and snow caused by climate change has opened the Arctic waters and wetlands to more deposits of mercury that can then be methylated.

There is yet another troubling amplifier. The most important driver of methylation is bacteria. The fact that climate change is creating a warmer environment where bacteria can thrive has the potential to enhance the proliferation of methyl mercury. "In all my years of working with mercury, the one thing that is the most important with regards to methyl mercury is you need those bacteria to produce it," St. Louis tells me. "So if climate change somehow affects their activity, that will have much greater impact in my opinion than adding more inorganic mercury to the system. Climate change can affect the bacteria activity by warming the water column [in oceans] and creating more wetlands in the north. Those wetlands in the Arctic can get quite hot because they absorb so much sun. So there will be more bacteria.

"We are only starting to understand it. The whole problem is with regards to what is going on in the oceans."

All of which leads us back to the frost flowers. Atmospheric mercury depletion events occur only in the spring and most powerfully over fresh sea ice along coastal regions. It is the formation of new ice that produces frost flowers laden with bromine. Those beautiful little ice crystals act as a delivery system for bromine to the atmosphere, helping to drive the chemical processes of atmospheric mercury depletion. They also act as a sink that captures the mercury and transfers it to the open ocean and ultimately into the food chain. As easily as my light touch shattered their lattice, winds blow them into open water. With more open water, more crystals will blow into the ocean, increasing the mercury content and the potential to create more methyl mercury. Climate change could lead to the production of more frost flowers as temperatures waver on each side of zero, creating a more rapid cycle of fresh ice—the ideal incubator for frost flowers—and open water. As it stands, scientists estimate that each year AMDEs are responsible for dumping up to three hundred metric tons of mercury over the Arctic.[6]

There is the possibility, however, that the warming of the Arctic could create a counterbalance to this process. St. Louis told me that climate change could lead to fewer atmospheric mercury depletion events because the atmosphere has to be below zero degrees Celsius for AMDEs to occur.

I pointed out that there always will be below-zero-Celsius conditions that create sea ice for these springtime atmospheric mercury depletion events, and asked why he thought there might be a counterbalance of fewer depletion events. His answer was simply that climate change could become so catastrophic in the Arctic that large areas of fresh sea ice might melt before they can release their saline brine into the atmosphere to create AMDEs. In this case, a large portion of benign mercury will remain in the air.

Concern over mercury poisoning of the food chain has led to restrictions on mercury emissions in North America and Europe, resulting in a reduction. But levels in Asia continue to increase. Climate change is altering atmospheric and ocean currents, bringing more mercury and other contaminants such as cadmium and lead into the Arctic. What's more, the eagerness to exploit Arctic oil and gas as well as other mineral resources will inevitably unleash a new crop of mercury into the ecosystem, not to mention a host of other noxious heavy metals.

I leave one day earlier than scheduled. Originally I was supposed to fly out on a CC-130 Hercules prop plane, but when the military reroutes the sluggish beast for a search and rescue operation, it looks as if I will be stuck in Alert for another five days or more. That doesn't suit my schedule at all. The concern on the faces of the soldiers and other civilians who are waiting to fly out with me is as evident as my own, and just adds to the general tension. They have been here a lot longer than I have. There is talk of placing a health watch on a few soldiers whose leave might be delayed. But we are saved when the military redeploys a CC-177 Globemaster ferrying supplies between Thule and Alert to take us back to Trenton. "They are bringing up the cargo pallet," one civilian tells me.

"What's that supposed to mean?"

"For our luggage." He adds that he's been here for two weeks. "I can take about a week and then I want out. If it weren't for the money, I wouldn't be here at all."

His instincts are good. I can hear the hum of the Globemaster's four jet engines as it approaches the runway. It is a fat but sleek military plane with a cargo hold that could swallow a Hercules whole. A small bus pulls up to the barracks to ferry us to the airport. Major Hoddinott and his staff clap as each name is called out. He then accompanies us to the airplane and shakes our hands as we board. Except for the seventeen passengers, the plane is empty. We strap ourselves into the canvas fold-up seats that line this windowless cigar tube. The two pilots waste little time. They steer the massive plane onto the frozen runway and immediately take off into a cloudy sky. We head due south to Thule to refuel. Here, the temperature is balmy and the Arctic waters are clear blue with a few icebergs floating on the horizon. To the east, several kilometers short of the runway, I see the long rounded slope of the Greenland ice sheet as it stretches towards the sea. You can see the telltale signs of its gradual recession in the scars it has carved on the bare rocks. Still, there is no mistaking its unbelievable breadth, filling the entire eastern horizon with a towering wall of brilliant white snow and ice. It is an elephant where Devon is a mouse.

Several hours later, we are back in the air. I sit up in the cockpit hoping to see the ice sheet from above. The cockpit gives a panoramic, picture-window view of the world, but unfortunately clouds shield most of the glacier. We follow a flight path over Baffin Bay, down the length of Baffin Island, over Hudson Strait, Ungava Bay, Québec and Ontario, landing in Trenton just before sundown, having consumed 120,000 pounds of fuel and having emitted 168 tons of carbon dioxide.[7]

CHAPTER 8

WHAT OIL?

IN WHICH WE PONDER WAYS TO MELT ARCTIC ICE TO GET
AT THE OIL. HOW ABOUT USING PLANELOADS OF SOOT?

SPREAD OUT ACROSS THE FLOOR IN FRONT OF US ARE A LARGE
geological map of the Arctic along with a few seismic survey maps.
Andrew Miall and I are down on our knees following the undulat-
ing lines that represent subsurface rock formations. He's mutter-
ing about upfolds and anticlines, fault traps and stratigraphic traps,
anything that might point to what we all hunger for—oil.

Miall is a veteran petroleum geologist and resident stoic at
the University of Toronto. He spent many of his early years surveying the Arctic for oil and gas in the late 1960s and early 1970s.
The reservoirs and wells he helped find were never exploited
because of ice and climate. Forty years later, the search is on again
and we're trying to separate fact from fiction, reality from hype.
"There undoubtedly is a lot there," Miall says of the Arctic. The
question is how much and where.

There are already more than four hundred known oil and
gas fields above the Arctic Circle, mostly in Russia and along the
north shore of Alaska. Total reserves in these fields are 40 billion
barrels of oil and 1,136 trillion cubic feet of natural gas. The news

from the exploration wells that creep over Russia's vast continental shelf, which in some areas extends as much as 1,700 kilometers above the Arctic Circle and also out into the Chukchi Sea off the northwest coast of Alaska, is even more promising.

Brash economics drives Arctic development despite the global threat of climate change. Since the 1970s, the Alaska economy has been heavily dependent on the North Slope oil fields that look out over the Beaufort Sea. With no income tax or sales tax, the state funds about 90 percent of its budget from royalties and taxes on oil and gas producers. Since 1981, the oil revenue–endowed Alaska Permanent Fund has cut an annual dividend check to every citizen in an amount that varies with the price of oil. The dividend peaked in 2008 at $2,069 per resident and fell in 2009 to $1,305. For a family of four, that's a substantial payout. In Alaska, oil is king.

While oil is the ultimate source of Alaska's wealth, the Trans-Alaska Pipeline is the facilitator. Without the pipeline, the oil would not get to market in as timely a fashion. Since 1977, oil from the North Slope fields has flowed through the pipeline 1,287 kilometers south to the ice-free port of Valdez. At its peak, the pipeline was flowing at a rate of 2.1 million barrels a day. Today it flows at about 647,000 barrels a day, and the decline rate averages 6 percent a year. Alaska supplies about 3.3 percent of America's needs. Alaska state senator Lesil McGuire, chairwoman of the Senate energy committee, says that if the flow rate falls below 500,000 barrels a day, the pipeline begins to run into trouble because the oil temperature dips below freezing. This causes ice buildup, which can crack underground portions of the pipeline. "That is only 145,000 barrels away," she says.

With Alaska's existing wells in decline, the state government believes its salvation lies offshore in the Chukchi Sea, where the United States Geological Survey estimates there are 25 billion barrels of oil. In 2008, the U.S. federal government sold $2.66 billion in oil and gas leases on the outer continental shelf of the

Chukchi Sea to Shell Oil, Statoil and ConocoPhillips. Shell Oil planned to be pumping more than one billion barrels a year out of its leases by 2017. But those plans are on hold after the federal government slapped a moratorium on development of offshore leases in the Arctic following the Gulf Coast disaster. Normally, Alaska wouldn't care. The owner of the offshore resource is the federal government; Alaska won't collect a cent in royalties or taxes. But Alaska, McGuire says, needs that oil to keep the pipeline working. Thirteen thousand direct jobs rely on oil flowing through the pipeline.

There is one other reason Alaska wants the Chukchi oil to flow. Under Alaska's North Slope lie an estimated 20 billion barrels of heavy crude oil, which belongs to Alaska. To get heavy oil to flow, you have to mix it with light crude. The Chukchi oil is ideal for heavy oil extraction, McGuire says. "The supply itself [of Chukchi oil] keeps that Trans-Alaska pipeline alive, keeps those jobs, keeps our state growing." Without the Chukchi oil, she says, by mid-century Alaska will be a dying if not a dead oil and gas state. "We look to the Arctic as our future. I believe Alaska will continue to play a major role in the global energy picture."

To McGuire, climate change is an issue, but only insofar as it might "disrupt polar bear patterns."

Norway, Russia and Canada are different from Alaska only in scale. Each of their economies has become so reliant on oil and gas production that any sudden or even gradual decline would create hardship and potential social upheaval. Of Canada's total exports to the United States in 2009, 54 percent were oil and gas.[1] Like Alaska, the Canadian government can't see its future beyond the oil patch.

Miall draws an imagined line with his finger along the continental shelves extending off the north shore of Siberia out into the Chukchi

Sea. All continental shelves have potentially petroleum-bearing sedimentary rocks and Russia has the broadest continental shelves of all the Arctic nations, making up about half the total. "There has been quite a bit of exploration on the edge of the shelf here by the Russians and it has all been very, very promising. There are gas fields already producing from the edges of the shelf and you know it is quite reasonable to project these geological features out. So the potential of that area is quite clearly great. Nobody would ever dispute that the ownership of that is quite clearly Russian."

In Canada and Greenland, oil exploration began with a frenzy decades ago and then suddenly stopped in the late 1980s as the ice and cold made it too expensive. The first well was drilled in Arctic Canada in 1961, and over the next twenty years oil companies found 1.6 billion barrels of oil and 31.2 trillion cubic feet of natural gas. Not big, but not insubstantial. The effort, however, was enormous. In some cases whole islands were built to support the rigs operating in shallow waters. In other cases, companies created floating ice platforms with half natural sea ice and half man-made ice to hold the 450-ton rigs. Billions were spent, much of it tax dollars, but only a fraction of the High Arctic oil ever got to market. The wells were ultimately shut down and plugged because of high costs and bad climate. As the ice retreats, these orphaned oil fields beckon.

How much still awaits discovery? The figure bandied around is about 90 billion barrels of oil. It is an estimate that was originally made on July 23, 2008, by the United States Geological Survey in a study called "Circum-Arctic Resource Appraisal: Estimates of Undiscovered Oil and Gas North of the Arctic Circle." The study claimed that in addition to the 90 billion barrels of oil there are 1,670 trillion cubic feet of technically recoverable natural gas, and 44 billion barrels of technically recoverable natural gas liquids in twenty-five geologically defined areas thought to have potential for petroleum. The study also said the Arctic "accounts for about

13 percent of the undiscovered oil, 30 percent of the undiscovered natural gas, and 20 percent of the undiscovered natural gas liquids in the world. About 84 percent of the estimated resources are expected to occur offshore." In a world running out of oil, this was important news, even though the 90 billion barrels represents only three years of current global consumption. The study naturally produced headlines around the world. Arctic countries licked their lips and looked to buttress their claims.

A closer look at the numbers, however, tells a different story. David Hughes, a retired petroleum geologist with the Geological Survey of Canada, analyzed the detailed USGS data for the geological region of Baffin Bay, between Greenland and Canada. Here the USGS estimated reserves totaling 7.2 billion barrels of oil. Hughes said in a paper, "There is a 95 percent probability that at least zero oil resources exist, a 50 percent probability that 0.26 billion barrels exist and a 5 percent probability that 34.5 billion barrels exist. So the USGS rolls the dice and reports a 'mean' estimate of 7.265 billion barrels to the media, when this amount has likely a one in ten chance or less of existing."

Miall, who is also a former president of the Academy of Science in Canada, says the USGS estimates often serve the oil industry's political needs and what he claims is America's myth of energy independence. "For an economy that's built entirely on oil and gas and the clients they serve and the industry, the more oil and gas you claim is out there, the greater the likelihood that the government will spend money to go up there and search for it."

In the end, however, does it really matter if the USGS is wildly off and a fairer estimate would be, say, only 50 billion barrels or 20 billion or even 10? Miall doesn't think so. "The fact of the matter is, at one point we will have to exploit everything that is left unless there is a breakthrough in physics and we discover an entirely new form of energy. We are going to need the energy. There is just no way around it."

———

Miall spent a good part of his early career looking for oil in the Canadian Arctic archipelago. As we continue to comb over the maps, he points out the structures.

An oil reservoir needs essentially three elements: a source of biomass, a so-called "kitchen" where the mass is cooked and turned into oil or gas, and a trap that basically pools the oil and gas into a reservoir, stopping it from being pushed by water up to the surface, where it would dissipate. Fossil fuels are formed from biomass that is cooked underground at about 70 degrees Celsius over hundreds of millions of years. The richest source is marine plankton. The fuel is pushed upwards by either geological activity or water or both. A reservoir forms when the petroleum is pushed against a trap that stops it from going to the surface. The best trap is a salt layer, formed millions of years ago by intense evaporation of huge bodies of seawater. "The salt itself is impermeable. Like plastic. If you get oil and gas pooling underneath, it makes a perfect seal." Folded rock formations and impermeable faults where rock layers push up against each other can also serve as traps. Saudi Arabia has unusually effective traps and a massive biomass source deposited in the sand about 200 million years ago, when Arabia was part of the Tethys Ocean. As Africa drifted north against Asia, the collision with what is now Europe and Asia pushed up the seabed to create Saudi Arabia. (The same action created the Alps and the Himalayas.) The biomass that had been deposited at the bottom of the Tethys Ocean over millions of years pooled in giant rock faults under Saudi Arabia, creating large traps in which the world's largest oil reservoirs formed.

Without all three factors, there is no chance of finding oil. Miall shows me the high-potential geological formations around the Mackenzie Delta. "There's all sorts of nice things going on there," he says, pointing to several rock traps on a seismic map.

The Arctic tectonic history is not well known, but the region doesn't have the same dramatic faults and thrusts found in the Middle East. In fact, in the Arctic's south-central and western islands, Miall says, the rock is flat-lined, meaning there is less potential for traps. In the eastern archipelago, the rock is too old. It's igneous rock, formed from lava, and there is no oil in that.

Miall was first sent up to the High Arctic to look for oil in 1969 while his boss, the legendary cowboy oilman John Campbell (Cam) Sproule, tried to persuade New York investors that he could melt the Arctic. Sproule was a gifted promoter and succeeded in raising hundreds of millions of dollars from government and private investors eager to explore the next frontier. He founded PanArctic Oils, which became a major Arctic player. Its main shareholder was the Canadian government. Its leases were in the western High Arctic islands—the so-called Sverdrup Basin—where it spent more than $900 million drilling more than 175 wells and discovered about 17.5 trillion cubic feet of natural gas. It also discovered oil, and from 1985 to 1996 shipped 2.8 million barrels by tanker to Montréal, stopping only when the price of oil collapsed. It was a trickle in global terms, but it served the purpose of helping to establish Canadian sovereignty in the region.

"Cam Sproule, he was a little bit of a rogue," Miall says. "For several years he made his money out of running ground surveys across areas that oil companies were interested in. There were a lot of speculative investments in landholdings in this area back in that time. There were American investors from New York, these big fat-cat guys, and they thought they were going to make a killing out of taking out oil and gas permits in the Arctic, hoping that . . . they could sell the drilling rights to one of the major oil companies. One of the conditions of taking the land permit was that you have got to actually do some work on it. So Sproule would send these field crews up into the Arctic for weeks and months at a time to do a local survey. I spent two summers, long, long, long summers, up there and was kept in the field far too long because he was making a per diem keeping us up there. We were stuck up there until late August, early September, when the snow started to fly and it was pretty questionable whether we would be able to get flown out."

Miall taps his finger on an area of Ellesmere Island covered in glaciers and then recounts how Sproule tried to persuade New York investors to give him millions of dollars to melt the glaciers and drill for oil in the area, even though it was igneous rock. "He suggested that they would charter these Hercules aircraft and fill them full of soot and they would fly out over these glaciers and shovel the soot out the back onto the edge of the ice cap and the soot would sit on the ice and of course absorb the sun's heat and melt the ice and then you could see what was underneath and actually do some exploration. I was in a meeting where he was seriously suggesting this. I couldn't believe my ears."

In the Canadian Arctic, there are five main areas of interest: the Mackenzie Delta and corridor, the Sverdrup Basin, Baffin Bay and the Ellesmere margin, Hudson Bay, and eastern Labrador. According to Donna Kirkwood, the director of the Atlantic and western branch of the Geological Survey of Canada, their mapping so far indicates that these regions hold reserves of 28 billion barrels of oil and about 97 trillion cubic feet of natural gas. "These are very, very conservative estimates," she says, based primarily on scientific assessments of "what we know about reservoir rocks and migration paths from source rocks to reservoir rocks." In other words, they are not based on drilling information other than what was obtained in the 1960s and 1970s. About 40 percent of the Canadian Arctic has not been mapped to modern standards and Kirkwood says the more they survey the Arctic, the higher their estimates rise. While some are considerably higher than those published in 2008 by the U.S. Geological Survey, others are not. A case in point is the Sverdrup Basin. Canada estimates it holds at least five billion barrels; the USGS estimates are less than one billion.

Canada says Baffin Bay and the Ellesmere ridge hold 11 billion; the United States agrees. The point is that before you start drilling, nobody knows the truth. But when the tide flows in favor of drilling, then drilling it will be.

The Sverdrup Basin, which experienced the most intensive drilling of any area in the Canadian High Arctic during the 1960s and 1970s, has the deepest sediment pile, over ten kilometers thick. Farther south, in the Mackenzie River Delta of the Yukon and Northwest Territories, just over one billion barrels of oil have been discovered. There is evidence that the Mackenzie is actually a young river geologically speaking, which reduces the likelihood of big oil finds.

Jack McMillan, the Canadian geologist who discovered the fossil forest in Hot Weather Creek on Ellesmere Island, speculated that a river system that flowed across the interior of northern Canada from the Rocky Mountains eastward deposited the thick quantities of sediment on the continental shelf off Labrador and probably off Baffin Island, which is why there could be substantial amounts of oil and gas there. In other words, the Mackenzie River flowed into Hudson Bay and not the Beaufort Sea, as it does today. The theory is that the ice age glaciers blocked this eastern flow and turned it north into what is now the Mackenzie River. Another geologist with the Geological Survey of Canada, Alejandra Duk-Rodkin, has since found ancient river channels and river terrace deposits that support McMillan's theory.

Between 1974 and 1983, oil companies drilled twenty-eight exploration wells into the sediment off Labrador and three wells off the southern half of Baffin Island. They found five large fields with an estimated 4.2 trillion cubic feet of natural gas. The wells were all shut in. They are smack in the middle of iceberg alley, where the giant icebergs calved from Greenland's and Canada's glaciers congregate. The shelf in that area is pretty shallow and icebergs can easily tear out underwater pipelines.

Miall says the final untapped area that promises to be among the richest fossil fuel sources in the Arctic is the eight-kilometer-wide Lancaster Sound, which serves as the eastern gateway to the Northwest Passage. Here a major fault line with sediment more than four kilometers deep could hold structures that have trapped large amounts of oil and gas. The trouble—if trouble is the word—is that these waters constitute one of the most ecologically rich areas in the Arctic. Lancaster Sound teems with wildlife and the Inuit have used it as a hunting ground for centuries, if not millennia. Narwhals, beluga whales, walruses and seals, along with many species of birds, migrate to these waters in the summer. About 40 percent of the world's belugas travel through the Sound to feed and give birth. This natural wildlife refuge has already created roadblocks not only for the exploitation of the immediate area but also for the exploitation of oil and gas all over the Canadian Arctic.

In 2008, the federal government gave the Geological Survey of Canada's Geo-mapping for Energy and Minerals (GEM) division a five-year, $100-million budget to explore for oil and gas as well as mineral resources in the Arctic. In the summer of 2010, GEM joined with the Alfred Wegener Institute for Polar and Marine Research in Germany to hire a German icebreaker, the RV *Polarstern*, to conduct seismic surveys of the underwater structures in Lancaster Sound. The Germans said they wanted to explore the tectonic plate history between Canada and Greenland. They also wanted to gather data relating to climate change research, such as methane concentration in the water column and water and ice temperatures, which of course doesn't require seismic equipment. The Canadian government maintained to the general public that they wanted to chart the boundaries of a $5-million marine park, which the Canadian environment minister at the time, Jim Prentice, had announced several months earlier. At the time, he called Lancaster Sound "one of the richest marine mammal areas in the world." The announcement coincided with the government's stated need to assert sovereignty over the Northwest Passage. When several Inuit communities in the region complained that the seismic surveys would threaten the health of the wildlife, another narrative emerged.

In 1987, the government had taken seismic surveys using dynamite, which the Inuit claimed had scared away the beluga whales for years afterwards. For the 2010 surveys, the scientists planned to use blasts from air guns, each one equivalent to about three thousand pounds per square inch of pressure, or 200 times standard air pressure. The *Polarstern* would tow an array of six to eight air guns floating horizontally six meters below the surface and firing off every fifteen to sixty seconds, generating a constant sound wave field strong enough to penetrate the earth's crust many kilometers down to the top of its mantle.[2] Modern ocean-bottom seismometers capture an incredibly clear and accurate picture of the deep rock formations. Detailed 3-D seismic is then used to zero in on potential oil and gas reservoirs within these formations.

GEM needed logistics support in Grise Fiord, where project director Gordon Oakey wanted to store a fuel supply for a helicopter and establish a field base. The Canadian government had obtained permission from the Nunavut government, but the hamlet councils of all the communities of Nunavut, including Grise Fiord, had not been informed.[3] When the government delivered two hundred barrels of Jet B aviation fuel to Grise Fiord on the summer sealift with no advance warning to the community, the elders demanded to know what was going on. Oakey was sent up in September to explain the situation. He apologized to the community for the "breakdown in communications"[4] and proceeded to explain what GEM was planning for the region. To a group of thirty Inuit hunters, he described the upcoming seismic surveys to be done off the *Polarstern* as well as the airborne magnetic and gravity surveys to map the fault systems of the sedimentary basins in Lancaster Sound. "I have been given the responsibility of identifying petroleum resource potential in eastern Nunavut . . . If naturally occurring oil seeps are identified, sampling will be carried out to retrieve rocks exposed at the seafloor," he told them.

When the Inuit expressed concern about potential damage to the caribou herds posed by land-based operations, Oakey told them that it was Inuit-owned land and in many areas they held the subsurface resource rights. "If there are petroleum resources there, they belong entirely to the Inuit," he said. They then raised concerns about the eventual impact of oil exploration on global warming. "Climate change is not my area of expertise," Oakey told them. "But having worked in the Arctic for many years, I have seen changes in the extremity of the seasons, thinner winter ice . . . and earlier breakup of sea-ice in the spring. I remember in 2001, when I was on the *Louis St. Laurent* cruise in Nares Strait, the two students from Grise Fiord made the comment 'I've never seen a mosquito before.'" When they questioned him about the impacts of the *Polarstern* seismics on marine life in Lancaster and Jones Sounds, he assured them that the seismic survey "has a significantly reduced impact on fish and marine animals compared with explosives."[5] The community didn't buy it.

In the spring of 2010, five Inuit villages on or near Lancaster Sound went to court to stop the survey. The day before the *Polarstern* was scheduled to start work, the court ruled in favor of the Inuit and ordered the surveying stopped on the basis that the Inuit had not been properly consulted about the possible harm the air guns could do to the wildlife upon which they relied for food. The court ruling was based on precedents set since 2004 by the Supreme Court of Canada. While the Canadian government has the right to exploit the natural resources of the Arctic waters, the right of consultation implies that the Inuit have a degree of veto if a project violates their hunting and cultural traditions, which of course are closely connected to the health of the land and its wildlife. It was not enough to gain the support of the government of Nunavut, whose base is 1,500 kilometers south of the contested area; the government also has to consult with each community council. The government had ignored them. The Inuit fought back and won. A new element of Arctic sovereignty had asserted itself.

Federal lawyers advised the government that it could ignore the court ruling and carry out the surveying because the waters in question are far enough from the shoreline to be federal territory. But the government decided it didn't want to antagonize the Inuit. Kirkwood confidently says the surveying will be done sometime in the next few years, but that will very likely depend on the Inuit hunters of Lancaster Sound and Grise Fiord.

Prime Minister Stephen Harper has claimed that unless Canada "uses" the Arctic, it will "lose the Arctic." It is a bogus claim, experts say. Over more than a century, Canada has surveyed and mapped the entire Arctic as well as supported communities throughout the region. No country questions Canada's claims to these lands. Harper went on to say: "We know from over a century of northern resource exploration that there is gas in the Beaufort, oil in the Eastern Arctic, and gold in the Yukon. There are diamonds in Nunavut and the Northwest Territories, and countless other precious resources buried under the ice, sea and tundra. But what we've found so far is merely the tip of the proverbial iceberg. Managed properly, Canada's share of this incredible endowment will fuel the prosperity of our country for generations. And geo-mapping will pave the way for the resource development of the future."

There is one other potential bonanza awaiting development in the Arctic. It is considered in the oil and gas business to be cutting-edge stuff. It's called methane gas hydrates. These are bits of methane locked inside water crystals, often referred to as "frozen heat." They are literally everywhere in both the ocean sediment and the deep permafrost. The trick is how to extract them. One of climate science's worst fears is that the melting of the permafrost will release billions of cubic feet of methane into the atmosphere. The gas hydrates that interest energy companies, however, reside hundreds of meters below the surface, inside the permafrost and ocean sediments.

Yannick Beaudoin is a Canadian geoscientist and specialist in gas hydrates. He works in Norway for the United Nations Environment Program researching the possible environmental impacts of exploiting this rich reserve of energy. "We are just trying to see how different sources of unconventional hydrocarbons could fit into a sustainable global energy mix," he says. "And again, sustainability is not just purely economic."

The methane gas hydrate resource contains about twice as much organic carbon as all other fossil fuel reserves on Earth—an estimated 20,000 trillion cubic meters of gas. To put this in perspective, gas consumption in the United States in 2010 was 671 billion cubic meters. So the hydrate resource could keep America humming along at its current consumption rate for the next 29,000 years. It's more than enough to replace all the coal-fired power plants in the world.

Canada has been researching gas hydrates for the last eight years. The estimates are still wildly imprecise. In the Sverdrup Basin, they range from 671 billion cubic feet to 21 trillion cubic feet. In the Mackenzie Delta and corridor, the range is 300 billion to 350 trillion, Kirkwood told a meeting of energy experts in Montréal in 2010.

The gas is found pretty well everywhere in the Arctic's deep subsurface permafrost, as well as in most ocean sediments throughout the world. Extracting the gas safely and economically is still a challenge, but only because not much work has been done in this field. There is no real economic pressure to drill for it. This means we have time to try to find methods of extraction that are environmentally safe. This has not been the case with any fossil fuel exploitation to date. The recent revolution in shale gas exploitation is a case in point. Environmental concerns arose only after the drilling started. "Shale gas went very quickly," Beaudoin says. "They went from discussion to commercialization. A lot of that was proximity to consumers. And infrastructure. We certainly have a different situation in the Arctic. It certainly gives us a fantastic opportunity to look at it as a clean slate."

There is so much money invested in conventional natural gas and in shale gas exploitation that there isn't much interest in gas hydrates, which is regarded as an energy source for the future. The exception is in countries such as Japan that depend heavily on foreign oil and gas. The Japanese are hoping that gas hydrates buried beneath Japan's continental shelves will help reduce their dependency on foreign supplies, and they are working with the Geological Survey of Canada in the Arctic to test extraction methods.

They first wanted to determine if conventional oil and gas drilling can extract the natural gas from the hydrates. They drilled two exploratory wells, both in the Northwest Territories. In the first well, the Japanese simply drilled down and waited for the gas to flow. They quickly ran into problems, as mud clogged up the works. The second well proved successful, but only after they used heat and compressed air to free the gas from the ice. They proved you can flow the gas, but the question remains as to whether it can be done economically. Canada and Alaska are hoping that gas hydrates can make Arctic communities, which now burn diesel fuel that has to be trucked or shipped in, energy independent.

For Beaudoin, however, the question is still whether it is environmentally safe. Will the drilling release more methane into the atmosphere? Most oil and gas wells have what they call fugitive emissions. They leak. Thousands of leaking gas hydrate wells would multiply the greenhouse gas problem, and accelerate global warming.

CHAPTER 9

PROPOSITION 23 AND THE MIRROR ON AMERICA

IN WHICH WE CONFRONT THE UNITED STATES OF DUH

Voices from Cyberspace:

tasherbean:
You would think that a ballot measure that was funded by two Texas oil companies and two extreme right wing billionaires would be dead on arrival in ANY state, but particularly here, in California. Of course it's being sold by the right, like everything else they try to sell, as "your taxes will go up!!" Pathetic.
3:52 PM September 10, 2010 from web

tramky:
California is essentially bankrupt, strangled by the environmental lobby and their counterparts that have dominated the California landscape—and laws—for far too long . . . It has been outrageous, and perhaps Prop 23 can help bring an end to it. The future of California depends on it.
6:36 AM September 11, 2010 from web

Of all the battles the fossil fuel industry has waged in the United States to stop legislation on climate change, none displayed its brute determination better than the war it financed in California over Proposition 23, the petition to kill climate change legislation.

Despite the top-priority status Obama gave to climate change legislation when he took office, the Senate's three attempts had gone down in flames. Fossil fuel interests succeeded in watering down these bills to such an extent that they even alienated potential backers. The last bill, sponsored by Senators John Kerry, Lindsey Graham and Joseph Lieberman, didn't even make it to the Senate floor. Oil, coal and gas companies had proved to be the masters of the U.S. Senate.

American environmentalists turned to California as their last real hope. California's Global Warming Solutions Act, passed in 2006, was gaining support in other states and therefore posed a major threat to polluters. The state's environmental legislation normally has impacts that go well beyond its borders. As America's most populous state with the eighth-largest economy in the world, California has a history of setting North American standards in environmental law, which is a bit odd given that its air is among the dirtiest in the nation. It is America's largest car market. If you want to sell cars in California, you have to meet that state's tighter regulations, so you might as well make those regulations universal. Countries such as Canada follow standards set by California.

So in 2006, when California enacted the Global Warming Solutions Act, the first climate change regulation in the United States, its standards threatened to migrate all over America and into Canada. Which is exactly what began to happen. Soon, seven western states and four Canadian provinces—British Columbia, Manitoba, Ontario and Québec—joined California in a cap-and-trade program and in writing regulations in a regional effort to cut emissions.

The act requires California to reduce its GHG emissions to 1990 levels by 2020. This constitutes a reduction of about 30 percent from business as usual. It also requires that by 2050 the state reduce its emissions 80 percent from 1990 levels.[1] This is what the science demands and it is what many countries in the European Union have committed to. Attached to this law are a host of regulations that will start the process of emission cuts in 2012. Any delay would reduce the state's ability to meet its targets. The act, known as Assembly Bill 32 (AB 32), represented an enormous challenge to the oil companies, particularly those with refineries in California. They viewed overturning AB 32 as the fight of their lives.

Their first political opportunity came with the state elections in 2010. In February of that year, several California conservative organizations, led by the powerful Howard Jarvis Taxpayers Association and its 200,000 members, began a campaign to kill the act. With money from Texas and California oil companies—including Occidental Petroleum Corporation, which had recently discovered new oil reserves in California—they paid US$2.2 million to National Petition Management Inc. to collect signatures from California voters to get what became known as Proposition 23 onto the ballot for the November 2, 2010, general elections. By May 2010, they had obtained the necessary 433,971 signatures and the race was on. As the Adam Smith Foundation of Missouri, a conservative advocacy group that contributed US$498,000 to the effort to kill the global warming law, said: "California's environmental regulations are among the strictest in the nation . . . Unfortunately for the rest of the nation, these poorly designed and heavy-handed regulations often work their way east and become the law of the land within a few short years."

The wording of the proposition was important. The drafters avoided language that would outright nullify the global warming act. Instead, they sought to make it practically impossible to implement the law's provisions. The proposition would stop any agency of the state government from implementing any laws or regulations having to do with greenhouse gases or renewable energy until the state's unemployment rate was 5.5 percent or less for four consecutive quarters. That was the unemployment rate when the law was enacted in 2006. By 2010, it had more than doubled to 12.5 percent. It had been at or below 5.5 percent only three times since 1976, when the state began keeping employment records. So if Proposition 23 passed, the possibility of implementing the global warming legislation was basically nil.

By making action contingent on employment, opponents of the law claimed the issue was not global warming but jobs. And every American could relate to that. As James Kellogg, a plumbing and pipefitting union representative, wrote in a column in the *San Francisco Chronicle*, his union supported Proposition 23 because it would "protect jobs and benefits and save California families billions of dollars in higher energy costs." He added that any green jobs would likely go to China, where they could produce wind turbines and solar panels cheaper than in California. Supporters labeled Proposition 23 "The California Jobs Initiative." Opponents called it "The Dirty Energy Proposition."[2] U.S. Senate majority leader Harry Reid, a Nevada Democrat who was doubtful the U.S. Senate would ever pass a climate change bill, called the battle over Proposition 23 "a seminal moment in U.S. climate legislation."

As soon as Proposition 23 won a spot on the November ballot, financial contributions to the California Jobs Initiative— the pro–Prop 23 side—began to pour in from oil companies. By the November election, 98 percent of the money had come from oil interests, the lion's share from two Texas companies and one from Kansas. Valero Energy Corporation of San Antonio, Texas, had two of its fourteen refineries in California. (It also has a refinery in Québec, where it carries on business under the Ultramar label.) Despite losses of US$352 million in 2009 and US$1 billion in 2008, the company managed to donate US$508,000 towards the petition campaign. Once Proposition 23 got on the ballot, Valero poured in another US$4.1 million. Another major contributor was Tesoro, also of San Antonio, Texas, and with two refineries in California. It shelled out US$2.1 million. Both of these companies refine primarily a cheaper sour and heavy crude that spews thousands of pounds of toxic pollutants into the air.[3] They are considered among California's top polluters.

"These two companies have fought climate change [legislation] in Washington and want to kill it here so it doesn't spread anywhere else," said Steve Maviglio, spokesperson for opponents of Proposition 23. "The two companies have taken billions in profits out of this state in the last ten years. They see that at risk in the development of our clean-energy economy and they certainly will be on the hook for millions of dollars in costs to clean up their pollution. They would rather spend the money on a ballot measure than cleaning up their mess."

The third-largest single contributor was Koch (pronounced "coke") Industries Inc., of Wichita, Kansas. It donated $1 million through its subsidiary, Flint Hills Resources, also of Wichita. But it was Koch's vast network of conservative advocacy groups that worried the opponents of Proposition 23 the most.

The company is owned by Charles and David Koch, brothers who rank eighteenth on the most recent international Forbes billionaires list with a combined personal fortune of $44 billion. Koch Industries, the second-largest private enterprise in the United States, had revenues of $100 billion in 2009 from a variety of businesses, including forestry products, ranching, oil refineries and pipelines that produce up to 800,000 barrels of oil a day. Their father was a founder in 1958 of the ultraconservative John Birch Society and the sons are libertarians who carry their conservative banner like warlords. In Obama's first year as U.S. president, the

Kochs spent US$20 million lobbying primarily on behalf of its oil and gas interests against emission control legislation and regulations of toxic chemicals such as dioxin. This was 36 percent more than they had spent on lobbying throughout the entire eight years of the Bush administration. In addition, they have spent millions supporting Republican politicians who oppose caps on emissions. One major recipient has been the Oklahoma climate change denier Senator Jim Inhofe.

For decades they have been leaders in financing conservative causes through a host of institutions, think tanks and foundations, many of which they created. The Kochs are the symbol of the old economy versus the new. They campaign against pretty well any regulation or law designed to reduce America's dependence on fossil fuels and transform its economy to clean energy. Meanwhile, their refineries and pipelines have been convicted of an impressive list of criminal and civil violations of environmental laws. According to a 2004 article by the Washington-based Center for Public Integrity,[4] these have involved more than three hundred oil spills from Koch pipelines, discharges over the legal limit of the cancer-causing organic chemical benzene from a Texas refinery, gas leaks and falsifying documents. In all cases, Koch Industries escaped with relatively small fines after the Bush administration took office in 2000.

Koch's Minnesota refinery is one of the largest importers of the high-carbon heavy crude from the Canadian tar sands. This business is threatened by low carbon fuel standards, which is why Koch has campaigned against them in the United States, particularly in California, Massachusetts and Oregon. It claims the low carbon standards would "cripple refiners that rely on heavy crude feedstocks" and would be "particularly devastating for refiners that use heavy Canadian crude."[5]

Yet at the same time, they have invested heavily in ethanol production, taking advantage of government subsidies in that industry.

Plumbing the depths of government for special favors and largesse is not something they do willingly. In fact, they claim that they are pulled into it kicking and screaming in order to compete in markets contaminated by political favoritism. As Charles Koch stated in a *Wall Street Journal* opinion piece,[6] the fault lies in the government's "market-distorting" programs:

> Too many businesses have successfully lobbied for special favors and treatment by seeking mandates for their products, subsidies (in the form of cash payments from the government), and regulations or tariffs to keep more efficient competitors at bay.
>
> Crony capitalism is much easier than competing in an open market. But it erodes our overall standard of living and stifles entrepreneurs by rewarding the politically favored rather than those who provide what consumers want.
>
> Because every other company in a given industry is accepting market-distorting programs, Koch companies have had little option but to do so as well, simply to remain competitive and help sustain our 50,000 U.S.-based jobs. However, even when such policies benefit us, we only support the policies that enhance true economic freedom.
>
> For example, because of government mandates, our refining business is essentially obligated to be in the ethanol business. We believe that ethanol—and every other product in the marketplace—should be required to compete on its own merits, without mandates, subsidies or protective tariffs. Such policies only increase the prices of those products, taxes and the cost of many other goods and services.

A Greenpeace report in 2010 called Koch Industries "a financial kingpin of climate science denial and clean energy opposition."

The brothers are the billionaires behind much of the climate change fearmongering.

Between 2005 and 2010, the Kochs spent US$47.3 million lobbying Washington against tax increases, social programs and environmental regulations.[7] But they work most of their magic through think tanks. Koch money and expertise is behind the ultraconservative Tea Party movement that seeks to dominate the far-right faction of the Republican Party. Since 1997, the Kochs have given more than US$48 million in grants to a tangled group of thirty-nine conservative organizations, which together are often referred to as the Kochtopus.[8] They describe themselves as research and educational institutions. The Koch-supported Heritage Foundation sums up its mission thus: "to formulate and promote conservative public policies based on the principles of free enterprise, limited government, individual freedom, traditional American values, and a strong national defense."[9] Among the most active are the Cato Institute (which famously declared that the 2008 market collapse was caused by too much government regulation), the Mercatus Center, Americans for Prosperity Foundation and FreedomWorks, all of which the Kochs founded and continue to help finance. Their reports, websites, blogs and papers serve as a platform for climate change deniers and skeptics, who often attack the credibility of climate change science, oppose controls on greenhouse gas emissions and try to discredit scientists working with the International Panel on Climate Change. In 2009, the Cato Institute ran full-page ads in the *Washington Post*, the *New York Times*, the *Chicago Tribune* and the *Los Angeles Times* claiming that the "case for alarm regarding climate change is grossly overstated." The ad claimed, "There has been no net global warming for over a decade." This, of course, is flatly contradicted by the scientific consensus.

In Canada, the Kochs have given $175,000 to the conservative Fraser Institute, which has published papers opposing government efforts to regulate emissions, claiming, wrongly, that "from a scientific standpoint there is a powerful case to be made that [they] are wholly unnecessary." Many other Fraser Institute papers raise doubts about the global scientific consensus on climate change. (The institute has also received $120,000 from ExxonMobil since 2003.)

Of all the Koch-funded groups, the Heartland Institute of Chicago stands out as among the most blatant and persistent publishers of climate-change skeptics. For example, the institute's publications have run stories that claim "Electric Cars Threaten Energy Independence," "CO_2 Curbs Would Be Devastating" and "Carbon Dioxide Restrictions Will Devastate Environmental Treasures." It holds an annual International Climate Conference that is a stage for global warming deniers.

In a 2001 statement, the Mercatus Center, which was set up by the Kochs at the publicly funded George Mason University in Arlington, Virginia, told Congress that global warming is "beneficial, occurring at night, in the winter, and at the poles." By 2009, the Mercatus admitted that man-made climate change poses dangers, but suggested that populations simply be moved out of areas adversely affected by climate change in lieu of emission reductions.[10]

The pinnacle of denier success came after unknown hackers stole thousands of personal emails from computers at the Climate Research Unit of the University of East Anglia in Britain. Many so-called "experts" appeared on Fox News, CNN, the BBC and other networks to claim that the emails proved climate scientists attached to the Intergovernmental Panel on Climate Change had invented, hidden or fudged data to prove that the climate is warming. Patrick Michaels of the Cato Institute spearheaded the attack, appearing on radio or television twenty-seven times to voice "climategate" allegations. Five independent investigations totally absolved the scientists of any wrongdoing and demonstrated that the allegations against them were based on an incorrect reading of several of the emails. Nevertheless, the Cato and Heartland institutes continued to make their allegations, even citing each other as sources. In August 2010, Michaels took part in a Fox News show entitled "The Green Swindle" where the announcer, Glenn Beck, stated that "the emails reveal a plot among the world's top climate scientists to hide the real inconvenient truth that the evidence supporting man-made global warming is far from conclusive." Michaels was quoted in a September 2010 publication of the Heartland Institute attacking the IPCC. The story stated, "Key scientists working with the IPCC hid, manipulated and destroyed scientific evidence that contradicted their alarmist claims." The story failed to mention that the scientists had been cleared.

Of all Michaels's television appearances, probably the most instructive was on Fareed Zakaria's CNN show on August 15, 2010. He appeared with Jeffrey Sachs, director of Columbia University's Earth Institute, and Gavin Schmidt, a NASA scientist. The discussion was about assessing the present danger of climate change and how to respond to it.

Schmidt: "We know that the planet is warming. This decade is the warmest decade that we have in the instrumental record. It's warmer than the nineties. The nineties are warmer than the eighties. The eighties are warmer than the seventies . . . And what we anticipate is that because we continue to add carbon dioxide to the system, we are going to continue to warm decade by decade by decade. The exact magnitude of where we are going to go is going to depend a little bit on the system but also on the decisions that we make as a society to either reduce carbon emissions or just carry on with business as usual."

Fareed Zakaria then asked Michaels if there was anything there that he disagreed with.

Michaels: "It's very clear that the planet is warmer than it was and that people have something to do with it. What we're concerned about is the magnitude and the rate of the warming . . . Simply saying that we should reduce emissions may actually be the wrong thing to do at the moment if you don't have the technology to really effectively do this and to do it globally."

Fareed asked: "What do you mean we can't do it effectively? We know how to reduce greenhouse gas emissions. We just stop using fuels that emit it. It may not be economically pleasant but that's different. We know how to do it."

Michaels replied: "We don't have a replacement technology. We simply don't have it."

Fareed: "I agree with that but that's different from saying we don't know how to do it."

Fareed then turned to Sachs and asked how, if we don't know how to wean ourselves off our fossil fuel–based way of life, we are going to solve the issue.

Sachs: "I think what Pat [Michaels] says is absolutely correct that you need a plan. But we need to get started now . . . I believe that the costs of inaction are so frightening for the world that they are beyond our imagining . . . They could be devastating for hundreds of millions of people. They could lead to war. They could lead to famine and that is not hyperbole. That is a very hard-headed assessment."

Fareed turned back to Michaels and asked: "When you hear all this doesn't it worry you? . . . Don't you want to do something about this?"

Michaels replied that what worries him are "opportunity costs." He explained that the climate change legislation that had gone before the U.S. Senate "would have cost a lot and been futile and when you take that away and when the government favors certain technologies and politicizes technologies you are doing worse than nothing. You are actually impairing your ability to respond in the long run. And that's my major concern with this issue . . . I'm advocating for efficiency."

Fareed: "Some people say that you are advocating for the current petroleum-based industry to stay as it is."

Michaels: "No."

Fareed: "And that a lot of your research is funded by these industries."

Michaels: "No."

Fareed: "Is your research funded by these industries?"

Michaels: "Not largely. The, uh, fact of the matter is . . ."

Fareed: "Can I ask you what percentage of your work is funded by the petroleum industry?"

Michaels: "I don't know. Forty percent. I don't know." Then he went on to say that technology will change "in a hundred years and we will very likely not be a fossil fuel–based economy in a hundred years."

Fareed: "You're confident we will be around in one hundred years."

Michaels: "Oh yeah."

Fareed then turned to Schmidt and asked what would happen if mankind did not take action to reduce greenhouse gas emissions.

Schmidt: "If we [allow business as usual to continue] we will end up with a different planet. We will end up with a planet that won't be recognizable in terms of where crops can be grown, that won't be recognizable in terms of where rain is falling, that won't be recognizable in terms of where glaciers are and ice sheets are and where the sea level is."

Sachs: "And to put that in human terms, that's a catastrophic planet. Not just a different planet . . . The ironic point is the combination of the technologies we have already in hand and those that are close on the horizon—if we do this sensibly we can do this at low cost, save the planet and save the economy."

Time was almost up. Michaels quickly interjected the last word: "And every time we threaten an apocalypse and it doesn't happen we cheapen the issue. Thank you."

Stunned silence. Michaels had come full circle, embodying the merry-go-round that has paralyzed debate in the United States.

The advocate for climate denial had now agreed on CNN that the planet is warming and we're responsible. The advocate for no government intervention was now claiming that all he wants is an efficient plan. Michaels was looking suspiciously like a liberal. But then he inserted his parting shot: "Every time we threaten an apocalypse and it doesn't happen we cheapen the issue." It was aimed at Sachs and Schmidt, implying that they are cheap doomsayers. Michaels was saying that we shouldn't warn people about the dangers of walking on thin ice because until someone actually falls through the ice you would be, in his words, "cheapening the issue" of the dangers of thin ice. Of course, this makes no sense. But when you are pouring oil on a burning planet, baffling the public about what is actually happening is the only weapon you have.

And Americans are baffled. While NASA and the National Oceanic and Atmospheric Administration join the IPCC in concluding that "warming of the climate system is unequivocal" and that there is "very high confidence that the global average net effect of human activities since 1750 has been one of warming," polls showed that by the fall of 2010 Americans had become increasingly doubtful. The Kochtopus seemed to be working.

In July 2010, the Koch brothers turned their guns on California and began pouring money into defeating AB 32. The arsenal was not just the Kochs' million-dollar donation. The Tea Party movement had given them an army of acolytes to serve the cause. The Koch-funded FreedomWorks sent in the youthful Brendan Steinhauser, its director of federal and state campaigns. Steinhauser was fresh out of the University of Texas, where he had published a book called *The Conservative Revolution* on how to organize conservative clubs on college campuses. Before coming to California to help organize the Tea Party, Steinhauser had attended the Heartland Institute's climate change conference in Chicago, where he learned the denial creed. "The basic arguments of most of the scientists at the conference are that climate change has always occurred, with cooling and warming trends; that it is probably due mostly to natural variability; that CO_2 has not made a major impact in the climate change trends; and that man has very, very little impact on the amount of CO_2 in the atmosphere," he wrote in his blog conservativerevolution.com. "These scientists have basically debunked the hoax of global warming alarmism, point by point. Every conservative activist should watch their speeches, read their books and articles and keep up with their blogs . . . We can win the propaganda war if we highlight our best minds in this battle and work to get them more exposure." In California, Steinhauser turned his talents to defeating the "left's lies about global warming," as he put it. He tried to rally his Tea Partiers behind Proposition 23, telling them to "bombard" Meg Whitman, the Republican candidate for governor, with letters urging her to support Proposition 23. Whitman opposed it.

Eight weeks before the election, the two sides appeared to be in a dead heat. A concerned California governor, Arnold Schwarzenegger, charged that Proposition 23 was motivated by "self-serving greed." Speaking at the fourth anniversary of AB 32, he said, "Does anybody really believe they are doing this out of the goodness of their black oil hearts—spending millions and millions of dollars to save jobs?" He called California "America's last hope for energy change."

The ballot box wasn't the only battlefield. There were a growing number of lawsuits or threats of lawsuits. The Howard Jarvis Taxpayers Association, joined by the California Chamber of Commerce, the California Small Business Alliance and the Western States Petroleum Association, sued the state over expenses incurred in implementing AB 32. The oil industry challenged California's low-carbon fuel standards.

In addition, Texas, Nebraska and Alabama threatened to sue California if Proposition 23 lost. They claimed that AB 32 violated constitutional rights relating to interstate commerce. In particular, they were talking about the flow of electricity from coal-fired power plants. As of this writing, North Dakota says it is considering suing Minnesota, claiming its 30-percent emission reduction law illegally restricts North Dakota's coal-fired power plants, which supply electricity to Minnesota. In another case, Alabama backed the Tennessee Valley Authority (TVA) in a lawsuit filed by North Carolina. TVA operates eleven coal-fired power plants in Tennessee, Alabama and Kentucky, which pollute North Carolina. North Carolina had won an injunction forcing TVA to install emission control scrubbers on four of the plants by 2013 to comply with North Carolina law. The Fourth U.S. Circuit Court of Appeals reversed that decision in July 2010, thereby restricting a state's ability to legislate emission reductions that might affect other states.

With five weeks to go before the elections, the supporters of Prop 23 had raised US$8.4 million and began rolling out the television ads. They featured your average middle-class California housewife staring into the camera with worried eyes and a handful of presumably unpaid bills: "I want to do my part on global warming," she says. "All Prop 23 says is let's wait until people are back to work and we can afford it."

The opposition to Prop 23 assembled a broad base of individual, corporate and institutional supporters who by October 2010 had contributed more than US$12.7 million, which was 50 percent higher than the amount their antagonists, the oil companies, had donated. In all, activists formed fourteen committees opposing Prop 23. Most of them worked through a central committee called "No on 23—Californians to Stop the Dirty Energy Proposition." Major contributors included the Union of Concerned Scientists, the Sierra Club, the National Wildlife Federation, ClimateWorks

Foundation, the National Audubon Society and the Natural Resources Defense Council of New York. Tom Steyer, an asset manager, cochaired the opposition and gave US$5 million to the cause. Movie director James Cameron, creator of the eco-movie *Avatar*, tossed in US$1 million.[11]

The "No on 23" ran ads targeting the Texas oil companies' funding of Prop 23. One ad opened with photos of wind turbines and workers installing solar panels before fading to images of belching oil refineries. A male voice said: "California is outlining a clean energy future, a growing work force of bright Californians to harness wind and solar power to move our state forward. But two Texas oil companies have a deceptive scheme to take us backwards. They're spending millions pushing Prop 23, which would kill clean energy standards, keep us addicted to costly polluting oil and threaten hundreds of thousands of California jobs. Stop the job-killing, dirty energy proposition. Vote no on 23."

As California debated Prop 23, it was once again experiencing record-high temperatures that threatened to cause another season of wildfires. The United States National Academies, the main scientific advisers to the nation,[12] published studies that showed wildfires in the western United States were up sixfold over the last thirty years. The Academies' models indicated that a one-degree-Celsius rise in temperatures increased the area of burn as much as 312 percent in the Sierran Steppe mixed forest regions of northwestern, northern and eastern California, 231 percent in the central dry steppe and 74 percent in the southern semidesert and desert.

Oil, gas and coal companies around the world keep pushing the envelope, lobbying politicians to take no action that will slow their growth. In most cases, they are the ones with access to government and they are the voices that governments listen to. Their trump cards are jobs and potential tax revenue.

In Canada, a country with vast potential clean-energy resources including the ideal combination of powerful and consistent wind, ample sunshine and hydro, fossil fuel interests have successfully thwarted effective emission reduction regulations. The Canadian lobby registry shows that lobbyists hired by these companies are among the largest special interest groups on Parliament Hill. They lobby not only to shape legislation in the industry's favor but also to ensure that those industries benefit from the billions of dollars in government grants being issued for clean-energy and emission reduction projects. Since 1996, 1,570 climate change lobbyists have pounded the halls of Parliament. Their client list has steadily increased since then from just 13 to 109. While some of these lobbyists represent environmental organizations, educational and health institutions and diverse industries, the largest single group represents the fossil fuel producers, who have employed 465 lobbyists since 1996. The registry shows that they have repeated access to the Prime Minister's Office and to the offices of his cabinet ministers.

Their pressure has been unrelenting. Stéphane Dion, environment minister in the Liberal government from 2004 to 2006, recalls, "It's almost daily. They say, 'What can we do? There is no technology possible [for emission cuts]. Can you exempt us? Please.' This kind of pressure [is] always, everywhere."

Most of it comes from Alberta, the base of Canada's tar sands industries. Although it is home to just 10 percent of Canada's 33 million people, Alberta is responsible for about 70 percent of the country's oil production. With its refineries, gas flaring, coal-fired power plants, and tar sands (also known as oil sands) processing, it spews out one-third of Canada's greenhouse gases. That figure is expected to rise to a high of 50 percent by 2025 as emissions from mining, oil and gas extraction and refining increase substantially.

Indeed, from 2004 to 2007 alone, Canada's emissions from mineral extraction increased 61 percent, "largely due to increased activity in the Alberta oil sands," according to Canada's 2008 *National Inventory Report* on greenhouse gases. Since 1990, emissions from Alberta's mining of the oil sands have increased more than 200 percent.[13] At the moment, the tar sands emit about 36 million metric tons of greenhouse gases annually into the atmosphere. This is almost equal to the emissions from Canada's 12.4 million cars. And it is just the beginning. Plans to double oil production by 2020 mean greenhouse gas emissions could more than triple to as high as 127 million metric tons a year, according to industry predictions. So when the international scientific consensus warns us that the world is teetering on the brink of catastrophic climate change, Canadians roll the dice betting that this isn't true. It's not a gamble intelligent people would make. And it's certainly not a gamble Canadians have any right to make.

Mark Rudolph, a twenty-year veteran lobbyist who represents Suncor and Shell Canada, suggested that lobbying in the provincial capital of Alberta is unnecessary since the government is entirely on the industry's side. Alberta, in effect, represents the single most powerful lobby in Ottawa, according to Rudolph. "The [Alberta] government . . . takes somewhat of the same point of view as the denialist companies," he told me. "And they basically say to the feds—despite the fact that they are political brethren and the majority of the [federal] cabinet is from Alberta—'Back off. This is our domain. Don't bother us.'"[14]

Not only has the oil industry's Ottawa lobbying won them a reprieve from greenhouse gas emission regulations, it has also earned them billions of dollars in current and promised grants to develop carbon capture and storage technology. In 2007, Canada's major oil companies, along with some other energy-related businesses, created the Integrated CO_2 Network[15] (IC02N) to lobby for "a proposed carbon capture and storage system for Canada." With offices at Suncor headquarters in Calgary, it employs six lobbyists, including three who had been senior policy advisers to Prime Minister Stephen Harper. The main aim of the network, which includes two major coal utilities, is to secure financing for carbon capture.

That lobbying has paid off. In October 2010, Harper joined Alberta premier Ed Stelmach in making the unproven technology of carbon storage a cornerstone of Canada's carbon reduction strategy. The leaders announced grants totaling $1.6 billion to finance two pilot projects to capture and store carbon from two coal-fired power plants owned by two members of the ICo2N, Shell Energy and TransAlta Corporation. ICo2N's website says carbon capture and storage could reduce Canada's CO_2 emissions by 20 million tons, or 2.6 percent. But the enormous cost of such a nationwide push—estimated at about $16 billion—obscures the true picture of how far Canada has to go in curbing its emissions. The effort would fail to capture even the carbon equivalent of a single tar sands project. In April 2012, TransAlta pulled out of the project claiming it made no economic sense. And with this retreat the cornerstone of Canada's GHG reduction program crumbled like an oatmeal cookie.

In the end, however, few Canadians appear to care much about climate change. In the federal election of May 2011, the only politician who discussed it was Green Party leader Elizabeth May. Her determination paid off. With the help of the Canadian environmental activist community, which blitzed her British Columbia riding with a telephone campaign pleading with local voters to send an environmentalist to Ottawa, she won the Green Party's first seat in Parliament. Across the country, however, the Green's share of the vote fell to 4 percent, from 6.8 percent in 2008.

Just ahead of Canada on the global per-capita emitter charts is Australia. It is the world's hottest and driest continent and has recently experienced record cycles of drought and flooding that climate models indicate will worsen. Projections show that the kind of temperature records which have brought yearlong droughts while flooding coastal areas will become normal seasonal weather patterns, implying extreme temperatures that would be unlivable. Droughts will be 10 to 40 percent more intense, as will flooding.

Yet, like Canada, Australia has been slow to act. The main reason is that coal is the primary source—81 percent—of the energy fueling its industries and homes. Coal is also the cornerstone of its foreign exchange earnings, representing about 20 percent of exports, or $55 billion. Coal power is responsible for 38 percent of Australia's greenhouse gases. Still, the country's production and use of coal continue to rise. Production is up about 100 percent since 1990 and is projected to double again by 2030.[16]

Ignoring the threat of climate change, over the last two decades Australia has purposely built up its coal industry and used the promise of cheap fuel and extensive bauxite reserves to entice heavy breathers such as energy-intensive aluminum refineries to its shores. As a result, its economy has become increasingly reliant on coal exports and these huge greenhouse gas emitters. Coal alone contributes annually more than AU$50 billion in export revenues, according to the Australian Coal Association. The four largest companies operating mines in Australia earned about US$6 billion in clear profit in 2009 and well over US$75 billion in revenue. Coal royalties make up about 10 percent of the revenues of the coal states of Queensland and New South Wales.

Despite its modest commitment to reduce greenhouse gases to a mere 5 percent below 2000 levels by 2020, Australia plans to double coal exports. Its coal exports already represent more than 750 million metric tons of carbon dioxide annually,[17] which is almost equivalent to all the emissions generated by Canada. About 80 percent of its exports go to China, Japan and other Asian countries to fuel their largely dirty plants and industries. To keep the exports flowing, the country has undertaken a rapid expansion of existing mines, resurrecting old mines and creating new ones.

Australia's coal economy wields enormous influence over government policy, not only through the mining companies themselves but also through unions and the heavy industries that rely on cheap coal. So far, Australia has done little to reduce its emissions. It's a country that recently built a new coal terminal and wharf a meter higher than normal to account for rising sea levels. Australia, like Canada, has chosen to build itself a monster and now it has to feed the beast.

In 2009, Australia proposed a mining or carbon tax that would help pay for the conversion of 20 percent of the country's energy to renewables by 2020. The coal companies threatened closure of mines and launched a AU$22-million campaign against the tax. Macarthur Coal chairman Keith DeLacy suggested Australia wait until the rest of the world imposes a mining tax so Australian coal will maintain its competitiveness. "Let's get the whole of the world working together on this," he said. As he knew only too well, there is little hope of that happening in the near future. The proposed mining tax has been watered down to such an extent that the industry has saved about $60.5 billion over ten years.[18] The industry then announced billions of dollars in expansion plans.

The intransigence of industry had by the middle of 2011 delayed indefinitely implementation of Australia's proposed cap-and-trade program, called the Carbon Pollution Reduction Scheme. If the program is ever implemented, the government has promised many companies financial compensation. According to a report by RiskMetrics Group for the Australian Conservation Foundation, assistance promised during the first five years of the scheme would include AU$2.8 billion for Rio Tinto, AU$1.7 billion for Alcoa and AU$1 billion for steelmaker Bluescope. The top twenty polluters will reap AU$11.7 billion.

The Australian Coal Association also wants money to compensate coal producers for any losses. These producers include

Anglo American Metallurgical Coal, whose chairman, Seamus French, claimed in a widely distributed company report that the trading scheme would cost Australia 126,000 jobs. "Australia is walking the plank," he wrote, but offered no proof.[19]

At the UNFCCC climate talks, Australia is often an eager participant, chairing various committees and working groups. At home, meanwhile, it continues to build its coal empire.

On November 2, 2010, Californians voted 61.4 percent against Prop 23 and thereby reaffirmed their support for the climate change legislation. Even central valley counties, which are traditionally conservative, opposed Prop 23, albeit by slim margins. Despite an economy deep in recession, Californians would not be diverted from putting the environment first. Thousands of individuals joined clean-energy companies and environmental NGOs to contribute almost US$30 million towards the ultimately successful campaign, more than doubling what the oil companies had spent. Big oil was outgunned and outmaneuvered. Its scare campaign was turned against it. California did not see its future buried in the rearview mirror. It looked forward towards clean energy. The oil companies would now have to spend their money to comply with the state's climate change law—or find a court to outlaw it.

The defeat of Prop 23 was timely. It served to build a powerful bipartisan fortress in a corner of the United States where a degree of climate change sanity ruled. With, as a result of the 2010 elections, the House of Representatives in Washington back in Republican hands and the emergence of the newly elected power bloc of Tea Party climate change deniers, any hope for progress on climate change would have had to be abandoned had the companies won in California.

Unfortunately, nothing is ever certain in America. Victory is never complete. On the same ballot as Prop 23 was a crack in

California's environmental shield through which a lesser-known and far more complex petition called Proposition 26 emerged. Critics called it the "sleeper law." Few voters even bothered to mark its ballot entry. Fewer still knew its repercussions. On the face of it, it seemed like an honest attempt to stop the state from imposing unjustified taxes. But its scope is worrying and its effects on environmental legislation potentially devastating. It forces the state to approve any new fee or levy by a two-thirds "super-majority" instead of just a 50 percent vote. The reality is that the law increases the difficulty of imposing fees on polluters.

Most of California's antipollution measures are financed by fees imposed on industry. So the impact of Prop 26 could be enormous as California attempts, under AB 32, to impose a fee structure designed to reduce greenhouse gas emissions. According to a study of Prop 26 by the University of California at Los Angeles' School of Law, the law will "make it harder to impose or revise fees to fund these [emission control] programs in the future."

To understand the true intentions of Prop 26, you simply have to look at who was behind it. Fossil fuel interests and the tobacco companies were the main donors.[20] Chevron Corporation alone contributed US$3.7 million—almost one-quarter of the US$17 million spent to support the new law. While activists successfully campaigned against Prop 23 by targeting two Texas oil companies, Chevron crept to victory in a flanking action.

Ground the oil industry lost in the failure of Prop 23 could be restored by Prop 26. For these billion-dollar companies, battling pollution laws is a reflex action where the instinctive maximizing of profit trumps moral integrity and the overall well-being of society. They too are caught in their own "remorseless working of things."

CHAPTER 10

DEAD ZONES

IN WHICH ICE LOSS CHOKES OFF OXYGEN TO THE WORLD'S LAKES

I OWN A SMALL PLOT OF LAND ALONG A PRIVATE ROAD IN QUÉBEC'S Eastern Townships. The house is set back on a hill that descends into a long, shallow bay off Lake Memphremagog, the largest lake in the region.

Over the last ten years—but particularly in the last five— there has been a noticeable change in climate. Despite the fact that we are in the northern end of the Allegheny Mountains, at a fairly high elevation, snow comes about a month later than it did in the 1990s and melts about a month earlier. Overall, the winters are milder than they used to be and temperatures tend to fluctuate wildly. Heavy snowfalls can be followed by rain. It's hard to maintain an outdoor hockey rink. Violent rainstorms are frequent in the spring and early summer and again in late fall. The lake has warmed several degrees above normal. The heavy storms drain nutrients out of the soil and into the lake, causing the growth of toxic blue algae over sections of the lake almost every summer. You don't want to get these toxins on your skin. They can cause a burning rash. This phenomenon was unheard of in the past.

Game fish in the local rivers have disappeared, leaving the waters to chub and carp, which need less oxygen, feed on pretty well anything and thrive in warmer waters. They are the rats of freshwater marine life.

The violent weather has brought down trees and damaged our narrow dirt road. Over the last five years my neighbors and I have had to invest more than $20,000 repairing the road surface and installing additional drainage pipes and ditches to save the steeper sections from total washout.

Our region is but a tiny fragment of the enormous Atlantic drainage basin, which includes the five Great Lakes and the massive St. Lawrence River. The basin houses the greatest collection of large and small lakes, rivers, streams and wetlands on the planet. It is the globe's largest concentration of free-flowing fresh water. As such, it has an enormous impact on weather, affecting wind speeds, temperatures and precipitation for a broad region of North America.

Like the millions of lakes and rivers that mark Canada's landscape, the Great Lakes are the handiwork of the mammoth glaciers of the last three ice ages. More than 90 percent of the water in these lakes is glacier melt. So vast are these bodies of water that they contain more than 20 percent of the world's fresh water. Like the glaciers that made them, what annually rejuvenates these lakes and what makes them so robust, fresh and alive despite human pollution is ice.

Until recently, scientists had not paid a great deal of attention to the effects of climate change on the Great Lakes. Oceans get most of the research money, and the lake research that has been done so far only scratches the surface of an issue that inevitably will have profound effects not only on the more than 60 million people who live near and around the lakes but also on those affected by the weather systems incubated in the Great Lakes region.

There is no question that the climate of the Atlantic basin has changed, altering these lakes and waterways in ways few scientists could have imagined even a few decades ago. Obvious to anybody familiar with these lakes and waterways are the changes in the cycle of water storage, evaporation, winds and precipitation. Nor do they need to be told by scientists that the lakes are undergoing serious changes in the nutrient cycles—including carbon, nitrogen and oxygen—upon which life in the lakes is based. You would have to be blind not to see it.

While my neighbors and I were struggling with the problems associated with milder winters and heavier storms, about fifteen hundred kilometers west of us in Bayfield, a small Wisconsin town on the southwest shore of Lake Superior, Neil Howk was grappling with the same issues.

A ranger with the Apostle Islands National Lakeshore Park, Neil had moved to Bayfield about twenty-five years earlier and over the years had watched as altered weather patterns—particularly warmer winters and more precipitation—seemed to result in fewer people visiting the park. He became interested in climate change, but wasn't sure how to document it.

The park is essentially a small archipelago of twenty-two islands lying in a cluster off Bayfield and serving as added protection for the town's deepwater harbor. Two of the islands have ice caves, which were a popular winter destination for tourists until a warmer climate often made the journey across the ice too hazardous. All of the islands are uninhabited except for the largest, Madeline Island. Two hundred people live there year-round. They rely on ferries to travel the four kilometers to and from the mainland. In the winter, they use an ice road. But there's a transition period in early winter and early spring when the ice is too thick for ferry traffic but still too thin for an ice road, which strands the Madeline Island residents. So the dates of the ferries' first and last runs inevitably earned a mention in the local papers as well as in the harbormaster logs.

In 2008, Neil's son Forrest was casting around for a science project. His teacher had encouraged the students to choose projects with local significance. His father suggested he research something to do with the local effects of global warming. This led Forrest to focus on the decline of lake ice. The question was how to document it in such a way that the project would have historical significance. Satellite data went back only a few years. Forrest wanted data going back a lot further than that.

During his summer vacation, Forrest worked on the ferries as a deckhand. He figured the ferry company would have records detailing its first and last runs of the open-water season. Forrest thought these records would provide a good long-term history of the freeze-up and break-up dates of the ice in Bayfield harbor. "I was able to get the ferry records, but they only went back to 1970," he says. Forrest wanted to go all the way back to when the town was founded in 1856. So he checked the *Bayfield County Press*. In its early days, Bayfield had been in competition with other Lake Superior ports and had used the local paper as a platform to brag about its long shipping season and navigation capabilities. "Its navigation was open longer than other ports such as Duluth," Forrest says. "So they always mentioned when the navigation season opened and closed." He also found in the Bayfield archives records of the harbor navigation season from 1857 to 1888, meticulously kept by a settler named Andrew Take.

Eventually, Forrest compiled a complete 150-year record of freeze-up and break-up dates, from 1857 to 2007. After accounting for the change from wooden hulls to steel, which after 1970 allowed the ferries to operate longer, his study showed that the length of the ice season off Bayview has been steadily shrinking.

With the help of professors Jay Austin and John Magnuson at the University of Wisconsin, Madison, Forrest constructed a data analysis which showed that the period of ice cover off Bayfield over the last 150 years has decreased by 45 days, or about 3.4 days per decade. The analysis also revealed that the shortening of the ice season is accelerating. The most dramatic decrease has come since 1975, during which time the ice season has shrunk by an average of 14.7 days each decade.

Forrest's study won third prize at the 2008 International Science and Engineering Fair in Atlanta, Georgia, and was published in 2009 in the *Journal of Great Lakes Research*. The study represents the kind of anecdotal data that tracks a historical record of the effects of global warming and that is available for anybody who bothers to look. Climate change, in other words, is all around us. It's not rocket science. We just have to open our eyes.

What Forrest discovered in the Bayfield ice records is replicated in similar types of data gathered from lakes and rivers in Russia, Sweden, Finland, Japan, Canada and elsewhere in the United States. They all point to an accelerating loss of ice. A study by Magnuson and thirteen other scientists[1] showed that thirty-eight of thirty-nine data sets dating back 150 years for lakes and rivers in Russia, Finland and Japan demonstrated an accelerating reduction in the ice season. One time set for the ice cover on Lake Constance on the Swiss-German border dates back to the ninth century. "The criterion for 'total ice cover' was a walk across the ice of the main basin to transport a Madonna figure between two churches: one in Germany, the other in Switzerland. The figure remained on one side of the lake until the next ice-covered winter, when it was possible to carry it back again." While that criterion may sound a little imprecise, it still reflects a trend and indicates "cooler winters from the 13th through the 16th centuries after which there is a decrease [in ice]."

A 2010 study[2] of surface temperatures of 167 large inland water bodies distributed worldwide and including the Great Lakes using data collected by satellites showed a definite warming trend from 1985 to 2010. The average rate of warming was 0.045 degrees Celsius per year. The highest annual rate was 0.10 degrees Celsius, found in the high latitudes of the northern hemisphere. With that kind of warming, it's no wonder Forrest's study shows a huge decline in lake ice particularly over the last thirty-five years. Because his study journeys all the way back to 1857, it reflects the gradual warming that has slowly occurred during the industrial age.

Forrest's anecdotal evidence of warming also mirrors physicist Jay Austin's studies on the water and air temperatures of the Great Lakes.[3] Austin uses an array of temperature metering devices distributed in buoys all over Lake Superior. His work shows that while global mean temperatures over the last forty years have increased by about 0.2 degrees Celsius each decade, the increase has been much higher around the Great Lakes. The reason is probably that they are far enough from the cooling effect of the oceans, Austin says. Lake Superior's surface water temperatures have increased about 1.21 degrees Celsius per decade since 1985, or about 2.5 degrees Celsius over all. Since 1906, the increase has been 3.5 percent. So most of it has been in the last twenty-five years. These numbers are generally reflected in all the Great Lakes.

The decline of ice on the lakes is, of course, not steady on a year-to-year basis. There are years when Lake Superior, for instance, is completely covered in ice and years when there is no ice at all. "So what is interesting to me is that relatively small changes in climatic conditions lead to large changes in the effective ice cover," Austin says. "So why is the system so sensitive in this sense? That's what I'm working on. These are really outstanding questions right now that people really don't know a whole lot about."

The higher temperatures have decreased the ice cover by 50 percent and are creating stronger winds, Austin says. Scientists believe that increasing water temperatures in the Great Lakes are destabilizing not only regional climates but also the lake's ability to circulate oxygen and therefore to maintain marine life.[4]

Lakes have a natural vertical circulation system that distributes oxygen from the upper surface waters to deeper layers. This process is at peak performance in the fall and early winter months, when the temperatures of the surface waters and the deeper waters are more uniform, allowing surface waters to mix with the denser deep waters. Warmer winters mean less ice cover and a longer warming season. As a consequence, surface waters are exposed for longer periods of the year to direct sun and winds and therefore absorb more heat. Two things happen as a result. A stronger barrier forms between the oxygen-rich warm upper layers of the lake and the denser cool lower layers. The surface waters heat up more than usual and do not mix with the cooler deeper waters. Therefore the deeper layers of the lake are not getting fresh injections of oxygen. Without fresh oxygen, the lower levels use up their supplies. Depleted of oxygen, they become what are known as "deep water dead zones" where marine life cannot survive. The greater the amount of warming in the upper layers, the longer these dead zones last and the greater the possibility of massive fish kills such as occurred in Lakes Erie and Michigan in 2001.[5]

The second thing that happens is the dead zones lead to the growth of toxic algae, which can contaminate drinking water. Excessive amounts of nitrates (mostly from fertilizers) and other pollutants can contribute to dead zones. Lake Erie, the shallowest of the Great Lakes and bordered by intensively farmed land, has been stricken by a massive dead zone problem for several decades. Another dead zone, covering about 1,300 square kilometers, has developed over the last fifty years in the lower St. Lawrence estuary, destroying cod stocks. Marine scientist Denis Gilbert of the Maurice Lamontagne Institute in Québec blames half the problem on excessive pollutants (mostly fertilizers) flowing down the river. The rest is caused by the shorter ice seasons and warmer waters linked to climate change. The degree to which this kind of stratification is occurring in the Great Lakes needs further study.

Using climate models, marine biologists predict that warmer lake waters could dramatically reduce the distribution of most cold water species of fish, such as arctic char and brook trout, in Canadian freshwater lakes and rivers while giving greater space to warm water species, such as smallmouth bass and carp.[6]

Studies carried out by Austin and other scientists at the Universities of Wisconsin and Minnesota show that Lake Superior's warmer waters have caused an increase in surface wind speeds of almost 5 percent per decade. Warmer temperatures, stronger winds, ice loss and higher levels of evaporation (studies show evaporation rates increased 30 percent between 1960 and 1980 alone)[7] have become permanent phenomena in the Great Lakes and are the subjects of intense scientific review. We need to know how warmer waters will affect marine life, water levels, precipitation and drought. We may be getting more rain, but the old saying still holds: "Rain makes hunger; snow makes bread." Rain is unpredictable and can easily destroy a crop. Snowmelt, on the other hand, supplies a ready flow of water.

Climate models predict climate change will result in lower lake levels and reduced water supplies. Studies over the last fifteen years show that all the lake levels are down. In fact, most of Canada's drainage basins show a negative water balance.[8] Harbors such as Montréal's have seen a steady and unusual decline in their water levels over the last few decades while their navigation seasons lengthen. Harbor officials don't need to wait for scientific proof. They see climate change in the scarred hulls of freighters just as Forrest Howk saw it in the ferry records of Bayfield.

CHAPTER 11

THE TEMPTATIONS OF THE MOON PALACE

IN WHICH WE FORGE A MARRIAGE OF CONVENIENCE AMID A GATHERING STORM

ONCE WE'RE OVER THE GULF OF MEXICO, THE WOMAN BESIDE me gets out of her seat, climbs over my legs, grabs a bag out of the overhead luggage compartment and heads off to the toilet. About fifteen minutes later she is back. She has shed her sweater and jeans, and is now wearing a pink tank top and black yoga pants. As she climbs back over me, the word BABE, printed in glittering silver across her rump, flashes before my eyes. She flops into her window seat and smooths dyed-black hair out of her eyes. Up to this point I have paid scant attention to her. But the BABE sparks my interest. I steal a glance and catch the telltale signs of her war against aging. The stretched skin, the excessive moisturizing, the sharpened jaw-line. No doubt there are hidden scars, but I can't see them; that would take closer inspection. She's got to be in her mid-forties, but she's traveling back to her twenties. And why not? Why leave well enough alone when you have at your fingertips the power of a time machine?

We begin our descent. There's the usual routine of garbage collection, seats in the upright position, belts fastened, wheels down, and then the slight shiver of a smooth landing. Earlier in the day we stepped out of a world of cold and into this tube. Four hours later we spill out into a world of warmth. The sunseekers quickly drain away into small white shuttle buses that chauffeur them to the various resort hotels scattered in palm clusters all along the coast. I'm booked into the Ocean Coral and Turquesa resort, south of Cancún on the Mayan Riviera, near the tiny fishing village of Puerto Morelos. Because I'm not a packaged tourist, I have to grab a cab.

Twenty minutes later, I'm in the Ocean Coral and Turquesa's lobby of soft beige Yucatán marble where a light sea breeze welcomes me. The eternal sound of ocean waves washes through the entrance. Birds call. It all seems like paradise until you find out it's a soundtrack piped through speakers hidden in fake rocks.

The reception clerk, dressed in a tidy tan uniform, takes my passport and credit card details and wraps a gold strap around my right wrist. For the next ten days this band is my ticket to unlimited meals, drinks and cocktails in the resort's eleven restaurants and bars, plus access to four swimming pools, a spa, and of course crystal white sand beaches and a turquoise sea. All for less than $800. Forty years ago, Cancún and the Mayan Riviera was a sandbar and mango swamp. Take away the cheap jet fuel that brings the tourists and this place would fold like a tent. The Mayans would retreat back to their villages in the Yucatán interior. The mangos would reclaim their territory. And the alluring photographs of beautiful women under swaying palm trees would disappear from the advertising landscape of the industrialized world.

It is here in this sun-splashed holiday haven that the world will try for the sixteenth time to come up with an international treaty ambitious enough to meet the increasingly terrifying challenges of climate change, such as stopping the sea from flooding my Mayan resort. The chosen meeting place is just up the road from my hotel. It's a five-star, peach-colored resort called the Moon Palace. Until now it's been known as a honeymoon haven.

———

The journey to Cancún began six months earlier in Bonn, Germany, home to the UNFCCC and a half-dozen other UN environmental agencies. It was here that delegates met for the first time since Copenhagen to begin repairing the wreckage and getting the talks back on track. At least that was the stated desire. Given the jagged pattern of advance and retreat as the game plays out, it's impossible to know the true intentions of each one of the 192 countries involved in the negotiations.

In Bonn, the charged atmosphere of Copenhagen had melted away, leaving behind exhaustion and confusion. Delegates arrived still wrapped in the fog of forgotten purpose. "The confusion is, where is this going in the end?" a senior European Union delegate told me. "Is this leading to a treaty, and what does it look like? What about the Kyoto Protocol? What is happening in the United States? I think what you see is that there is a big realization that very little has happened in terms of shifting positions of countries since Copenhagen. Positions haven't changed. It is funny, because you still hear on the floor people saying, 'Oh yeah, a legally binding agreement.' Then at the same time you also hear people saying in bilateral discussions, 'It's not going to happen.'"

A year earlier the narrative had been the epic march to Copenhagen, where world leaders would finally construct the deal that would rally the world around a single global treaty powerful enough to steer us clear of the risks posed by man-made climate change. For whatever reason—chronic distrust, arrogance, selfishness—that story became a tragicomedy. The political failure of Copenhagen put the entire multilateral process on trial. For this reason the Cancún meeting was pivotal. Without substantial progress, the nascent carbon markets would collapse. The cure for a global ailment would then fall to individual countries, cities and regions, whose success would be uneven at best. Meanwhile, with each new day, the climate picture worsened on all fronts. Time was running out.

Signals coming from the United States House and Senate were increasingly negative, confirming the Copenhagen warnings of Senator James Mountain (Jim) Inhofe that America would never sign a climate treaty. Too many American lawmakers appeared almost giddy with delight at the failure of Copenhagen.

For many developing and poor countries, the problem is American intransigence. "It's that sort of debate in which the U.S. is saying, 'We will comply the way we decide to comply. We will do whatever we decide to do whichever way we decide to do it,'" a vice-chairman of the Kyoto Protocol working group said.

I asked him in Bonn what would happen if America was taken out of the equation—if America was simply expelled from the talks.

"Then the whole thing would be much easier, of course. Then you wouldn't even need two tracks, given that the Americans are not part of Kyoto."

As talks continued in Bonn and then moved to Tianjin, China, and finally to Cancún, the topic of the United States and what the talks would be like without the U.S.A. at the table was openly discussed in the corridors. The belief was that the Americans, knowing they could never get a climate change treaty through the Senate, were too constricted by domestic politics to negotiate anything more than the lowest common denominator; in a consensus process, they were leading everyone in a race to the bottom. The hallway strategy was that if the Americans were isolated, they would come back to the table with more agreeable proposals.

The United States, however, had gone to ground. Gone was the exceptional bravado of 2009, when Todd Stern and Hillary Clinton had wagged their fingers at everyone and blamed China for Copenhagen's failure—an attack strategy that proved remarkably successful. That game plan, however, would work only once. Faith in the United States' ability to back up its words with domestic legislation had been lost. So the Americans turned silent and invisible. In the run-up to Cancún, Stern was nowhere to be seen and his second, Jonathan Pershing, was mute. It was hard if not impossible to take aim at them when nobody could find them. Behind the scenes, the United States worked closely with the Mexicans to make Cancún the conference of the unambitious.

But the United States was not the only hurdle. Its fellow members of the so-called "Umbrella Group"—Australia, Canada, Iceland, Japan, New Zealand, Norway, the Russian Federation and Ukraine—were equally stubborn in their lack of commitment to significant GHG reductions. In all cases their stated priority was maintaining strong economic growth. Their foot-dragging was no doubt anchored in the belief that they are among the countries where climate change will have the least-negative effects, although that is strictly relative.

The peer-reviewed Climate Vulnerability Monitor 2010,[1] released just prior to Cancún, reinforced this belief. The monitor rates a country's vulnerability to the four main impacts of climate change: health, weather, habitation and the economy. The vulnerability rating ranges from low, which is graphically represented by a small green dot, to acute, a large blood-red dot. Moderate to severe vulnerabilities are shaded in various hues of yellow.

Pages devoted to African countries are blood-red. The same is true of most of Asia. Pages devoted to Europe, Australia, New Zealand and North America run green and light yellow, indicating moderate to low negative impacts. Australia may suffer increased drought—already a problem—and a degree of habitat loss, but the dots never turn red and negative impacts won't upset its economy until after 2030. There is a similar prognosis for Canada, Norway, New Zealand, Japan and Ukraine. Of the Umbrella Group, only the United States is headed into blood-red territory. The study projects it will suffer acute habitat loss after 2030. Low-lying states such as Florida, Georgia and the Carolinas will be flooded, as will their coastal communities. Desertification will spread in the West.

In other words, at least until 2030, these umbrella countries may be inconvenienced by climate change to various degrees, but they believe their wealth will allow them to adapt to the changes in order to reduce their impact.[2] In contrast, almost all of Africa, most of Asia and the Asia-Pacific region, as well as all small island states will suffer dramatic consequences. These countries have neither the financial muscle nor the technology to deal with the impacts of climate change, and by 2030 large areas will begin to become unlivable. About thirty to forty states of the United Nations will disappear. This picture, however, does not move the Umbrella Group. Faced with a danger that appears remote, it is easy to be complacent.

Canada, in particular, speaks of climate change as if it's a good thing—bringing it longer growing seasons and opening its northern regions to agriculture and resource exploitation. Going into Cancún, Canada won the temperature rise jackpot. The World Meteorological Organization said that over the past few years the country had experienced the highest jump in temperatures—on average three degrees Celsius—in the world. Mean temperature rises of three degrees Celsius or more above normal were found throughout the eastern Canadian Arctic and subarctic.[3] "Temperatures averaged over Canada have been the highest on record," the WMO stated.

Canada did not bother to send an official climate ambassador to Bonn after its chief negotiator, Michael Martin, was promoted to deputy secretary to the cabinet. Not until early autumn did Canada appoint his replacement. The country also refused to donate money to the Copenhagen Accord fast-track financing to help poor and vulnerable nations adapt to climate change. It said it would only loan money to the fund.

Meanwhile, Canada's diplomatic efforts on climate change were geared primarily to lobbying foreign governments not to impose environmental barriers that would hurt Canada's tar sands exports. These included joining an attempt—unsuccessful—to kill California's low-carbon fuel standard and to repeal U.S. restrictions on the use of dirty fuels by the U.S. military and other government agencies. Canada also campaigned—again unsuccessfully—against Europe's Fuel Quality Directive, which is designed to reduce emissions by promoting the burning of cleaner fuels and thereby help the continent reach its 20 percent reduction goal by 2020.[4]

Then, one month before the Cancún negotiations, Prime Minister Stephen Harper allowed an unelected Senate to kill a private member's climate change bill that had been approved by Canada's House of Commons on its third reading. The bill would have required Canada to reduce its greenhouse gas emissions to at least 25 percent below 1990 levels by 2020, and set a 2050 target of 80 percent. In other words, it required that Canada shape its legislation according to the science of climate change. The voice of the people had spoken. The voice of the appointed Conservative senators silenced it.

Canada, like Australia, had become delusional. Its path to this psychiatric disorder was particularly tragic because its vision had once been lucid.

In April 1990, Canada's Parliament held a series of hearings into climate change during which prominent scientists and senior civil servants testified about the dangers of inaction. Dr. Digby McLaren, a geologist and former director of the Geological Survey of Canada as well as president of the Royal Society of Canada and co-author of *Planet Under Stress: The Challenge of Global Change*, argued that mankind had become so powerful a biological force that it was destabilizing the planet. He told politicians that human activity was compromising the earth's life-support system. "Such behavior surely implies an incapacity to recognize that we live inside a sealed room with limited air and limited resources." Many other scientists, economists and technical experts discussed related issues, such as sustainable development. They testified, for instance, that it would be easy to reduce Canada's emissions by 20 percent of 1990 levels by 2005 and still maintain a robust economy, as one of the government's own studies had already demonstrated. The debate ultimately questioned the too-often unquestioned belief that the absence of economic growth means collapse. The fact that this discussion occurred at all and in a parliamentary forum was truly historical. It demonstrated a political will to address the pressing issue of global pollution and climate change. Out of these discussions, Canada and its provinces began not only to develop new environmental policies but also to contemplate new ways of managing its economy.

In the background, however, was the emergence of the tar sands development into a dominant economic force and the beginning of an equally powerful and fiercely uncompromising political constituency designed to protect its growth. Within ten years this constituency ruled Alberta politics. It soon took control of the federal Progressive Conservative Party, changing its name to simply the Conservative Party. Within sixteen years it had seized power in Ottawa and controlled Canadian politics. It had all the hallmarks of far-right conservatism. But its Conservative Party affiliation was a mere convenience. In the end, it was simply the politics of the tar sands. By Cancún, any hope of a new, cleaner world had long since vanished from the Canadian political landscape. Canada had entered an age of denial.

U.S. ambassador David Jacobson spotted it in a meeting he had on November 5, 2009, with Canada's then environment minister, Jim Prentice. In a subsequent cable to Washington, Jacobson appeared to make fun of Prentice, intimating that he was delusional. He noted that Prentice expressed surprise at the international opposition to the tar sands, admitting that his government "failed to grasp the magnitude of the situation." Prentice then talked about his "love for the outdoors" and how he considered himself "conservationist minded."[5]

As Cancún approached, the tar sands' political constituency was international. The sands had become an integral part of U.S. energy policy in the short and long term. British, French, Italian, Chinese and Norwegian oil companies had invested or were investing billions of dollars in the sands. Canada had like-minded allies.

. The same national and international political base had emerged to support Australian coal. Despite increasing homeland evidence of catastrophic climate change, Australian politics remained in a state of denial. Four months before Cancún, a Royal Commission in the state of Victoria issued a report on the bushfires that on February 7, 2009, had caused "one of Australia's worst natural disasters."[6] For the first time on record, temperatures in Melbourne had been above 43 degrees Celsius for three consecutive days, peaking at 46.4 degrees. More than three hundred grass and forest fires broke out across the state. Fanned by a fierce windstorm, they destroyed whole communities and killed 173 people, many of whom died trying to save their homes. The commission estimated the final damage at about AU$4 billion. The report warned that Australians should expect more such events. "It would be a mistake to treat Black Saturday [when four hundred bushfires were recorded] as a 'one-off' event. With populations at the rural-urban interface growing and the *impact of climate change*, the risks associated with bushfire are likely to increase."[7]

Once again, Australia's government took no action. Its emissions had increased 31.4 percent over 1990 levels by 2008—well above its Kyoto commitment of an 8 percent increase.[8] Its Copenhagen Accord 2020 target, which is not binding, was a meager 5 percent reduction on 2000 levels, with a promise to increase its reductions to as much as 25 percent if other nations made equivalent commitments. "The world is acting on climate change, with over thirty countries including the major nations of the European Union and Japan operating or implementing emissions trading schemes like Australia's Carbon Pollution Reduction Scheme," Senator Penny Wong said when the country announced its target in January 2010. Her words were pure fantasy. The proposed reduction scheme has never been implemented. (Neither has Japan's.) In other words, Australia had no plan to reduce its emissions. Happy that its role as spokesperson for the Umbrella Group lent it a major voice in international climate talks, Australia remained obsessed with expanding coal exports to China. Australian coal would continue to help drive the engines of China's export economy, which is primarily responsible for the accelerating growth in global greenhouse gases.

Australia soon paid the price for its conceit. After the Cancún conference, record-breaking torrential rains struck northeastern Australia, flooding an area the size of Germany and France combined, including the regions' coal mines, whose freight lines were cut off. Queensland state treasurer Andrew Fraser called the economic losses a "disaster of biblical proportions." The government estimated the damages at 0.5 percent of the nation's GDP. However, a ten-year drought with record-high temperatures and bushfires, followed by catastrophic floods, taught Australian politicians nothing about the risks of climate change.

Elsewhere in the world, resource-based countries were equally blithe. Middle East oil states, despite flash floods, eroding coastlines, scorching temperatures, blinding sandstorms and increasing water scarcity, held firm on their opposition to a climate change agreement.

As Cancún approached, the great march of countries pretending to seek a global solution to climate change continued. Only Europe seemed to have held on to its determination to forge a transition to a clean economy and lead the world in emission reductions. It was the only large emitter to translate its 20 percent Copenhagen commitment into law, and it claimed it had the numbers to show that a country could meet its targets and still enjoy economic growth.

The recession of 2008 had reduced demand for energy in Europe, and had led to an 11.8 percent drop in 2009 emissions, making the EU's 20 percent target by 2020 less onerous.[9] Still, by the EU's own admission, that target was not high enough to keep the global mean temperature rise below 2 degrees Celsius.[10] Europe was determined to reach 30 percent by 2020, though its business community opposed this higher target unless the Americans agreed to similar cuts, a faint hope if ever there was one. The European Union's reports showed that to keep temperature rises below two degrees, the developed countries would have to reduce emissions up to 90 percent from 1990 levels by 2050. Developing countries— primarily India and China—would have to commit to at least a 15 percent reduction from business as usual. So a European target of at least 30 percent below 1990 levels by 2020 was an absolute necessity if Europe was to lead the way.

Europe had assessed the cost and discovered it was surprisingly low. The total cost of a 30 percent reduction was 0.54 percent of GDP.[11] This amounted to about 70 billion euros per annum by 2020, a small fraction of the estimated US$3.1 trillion (or 5.5 percent of global GDP)[12] twenty countries spent rescuing themselves from the 2008 economic collapse. Increasing the target to 30 percent would amount to an annual cost equal to about 0.2 percent of the EU GDP by 2020. The EU estimates that the conversion to clean energy would save 40 billion euros by 2020 in reduced oil and gas imports. Health benefits and reduced costs in air quality pollution equipment would amount to savings of another 6.5 to 11.5 billion euros. According to the International Energy Agency, each year of delay in investment in clean energy will increase the global cost of emission reductions by about US$500 billion.[13] What's more, as climate change harms forests (primarily through storms and insect infestations), the likelihood that they will be able to bear a heavier burden as carbon sinks seems less plausible.

The EU's main concern about adopting a 30 percent target by 2020 is what it calls "carbon leakage," where companies take advantage of low-cost production in countries that have weak clean-energy laws to undercut European products. To counter this, the EU is considering tariffs or other instruments to increase the cost of imports and level the playing field. The EU believes this will force other countries to adopt low-carbon policies equivalent to the EU's. It's a card the EU intended to play at Cancún.

Entering into the Cancún negotiations, the recession in Europe still posed economic challenges to its low-carbon programs because of reduced corporate profits and diversions of money to bail out member states. The 85 billion euros the EU had just paid to Ireland was a case in point. Yet European countries generally refused to be diverted from their joint and national goals of clean economies, and believed that the transition to green technologies would be the main driver of global economic growth. It was determined to be the leader.

Great Britain's Climate Change Act, for example, sets a reduction target of 34 percent below 1990 levels by 2020. It plans an 80 percent reduction by 2050 and has created a new Department of Energy and Climate Change to manage the transition to a clean economy. Most of the effort will be in moving away from coal-fired power plants and erecting offshore windmills. "It's something the science is telling us to do and it's also frankly what our economic interests are telling us to do, because we want to develop these industries very rapidly," Chris Huhne, the U.K.'s secretary of energy and climate change, told me in Cancún.

Offsets, however, play a big part in Europe's emission reduction plans. This form of carbon bartering allows industrialized countries to finance cheaper emission reduction projects in developing countries under the Kyoto Protocol's Clean Development

Mechanism (CDM). The industrialized countries can then offset the alleged saving in GHGs against their emission targets at home. Most of this money has gone to projects in China, India, Brazil, Indonesia and South Africa, where emissions nevertheless continue to rise. Europe, however, can claim that it is cutting its emissions. Emission trading schemes are estimated to be responsible for up to half of Europe's claimed carbon reductions.

On a multitude of fronts, loopholes and inefficiency have corrupted the system. According to a study by Germany's Oko Institute, since 2005 Germany's nuclear power companies have made 39 billion euros in windfall profits by trading carbon credits that have been allocated to them free of charge.[14]

Also of questionable efficiency are the several billion euros that have been spent on CDM projects allowing developed countries to invest in climate-friendly projects in developing countries and claim the emission reductions as their own. More than a billion dollars was invested in CDM projects to reduce the greenhouse gas HFC 23. An EU study shows that it would have cost only about $100 million in direct investment to capture and destroy these gases in all of these projects.

"The CDMs are perverted," Jo Leinen, president of the European Parliament's Committee on the Environment, Public Health and Food Safety, told me. He also noted that there are 11 billion tons of offset certificates—called Assigned Amount Units—held by countries such as Russia, Ukraine, Poland and the Czech Republic that have not been used. This is equivalent to a third of the world's annual GHG emissions. These offsets have the impact of lowering the price of carbon credits. And if they are cashed in, they will give the false impression of an 11-billion-ton reduction in GHG emissions.

Europe's efforts, however, cannot be dismissed out of hand because of verification problems and loopholes in emission trading schemes, which it has plans to rectify. By Cancún, Europe was

moving forward with or without—and it was mostly without—
the rest of the world. Meanwhile, the rest of the industrialized
and emerging world was refusing to acknowledge the danger
ahead in any real terms. This threatened to render all of Europe's
efforts useless.

Eighteen years after the UNFCCC was signed in Rio de Janeiro,
negotiators were facing the unpleasant truth that all the pledges
made so far to reduce greenhouse gas emissions had not come
anywhere near to meeting the 2-degrees-Celsius goal. In fact, the
targets pledged in the Copenhagen Accord put the world on a
path to at least 3.5 degrees Celsius according to the International
Energy Agency.

At Cancún, the United Nations Environment Program
(UNEP) issued a report stating that humanity's inaction means
there is a 40-percent gap between emission reduction pledges and
the 2-degrees-Celsius target. That means we must cut another six
billion metric tons of carbon each year by 2020, which is more
than the total emissions of all the world's cars, buses and trucks in
2005. But, says the UNEP hopefully, "tackling climate change is
still manageable if leadership is shown."

Not only were the Copenhagen Accord's emission reduction
pledges proving empty, but also the promised US$30 billion in
fast-track financing to help the poorest and most vulnerable countries adapt to climate change had pretty well vanished. The money
was supposed to be spent between 2010 and 2012. But by Cancún
no mechanism had been worked out to distribute the money and
in fact none of it had been handed out in 2010. As much as 50
percent was now in the form of loans and not the grants that had
been promised. Industrialized countries, pleading poverty, had
refused to commit to a third of the total, and the rest was to be
diverted from other aid programs.

The fast-track financing had been a carrot designed to lure developing countries into signing an agreement they otherwise would not have accepted. The fact that by the end of 2010 the money had still not shown up threatened to erode whatever was left of the trust between rich and poor countries.

"So far the fast-start finance has neither been fast nor has it started, and there has hardly been any finance," Jairam Ramesh, India's environment minister, said in the first week of negotiations in Cancún. He added a note of warning: "Fast-track finance was an essential part of the Copenhagen Accord. If the current lackadaisical approach to fast-start financing continues, you are not going to create the conditions conducive to cooperative action. These negotiations are built on trust. There is a climate for negotiations and that climate does get eviscerated when one of the foundations of the accord, which is fast-start finance, is simply not visible."

Bad faith continued to corrupt the negotiating process. Trust was thin on the ground. But one fact grabbed everyone's attention: Even though the recession had caused a slight drop in energy consumption and emissions in Western industrialized countries, global emissions had continued to rise, largely because of India and China. Nearing the end of 2010, emissions were on track to reach record levels. The atmospheric space was becoming a tighter fit. Two years earlier, the atmospheric CO_2 content had reached slightly more than 384 parts per million. When the Cancún conference opened at the beginning of December 2010, it was nearing 390. The CO_2 molecules that keep our world in the Goldilocks zone were no longer merely creeping up the numerical scale; they were beginning a rapid climb and were now 40 percent higher than preindustrial levels (arbitrarily set at the year 1750).[15] Equally worrying was the sudden rise in methane molecules in the atmosphere due to permafrost melting in the Arctic. The North Pole was releasing its methane content in ever-increasing amounts.

These atmospheric GHG increases corresponded with a record global mean temperature increase in 2010,[16] which, according to the United Nations' World Meteorological Organization, was one of the three warmest years recorded since 1850. The ten-year period beginning in 2001 was the warmest since the beginning of instrumental records. The decade was about half a degree Celsius higher than the annual average between 1961 and 1990, which is an extraordinary hike. Corresponding to this temperature rise were record numbers of extreme weather events.

Massive floods swamped Pakistan; murderous mudslides swallowed villagers in China; unseasonal monsoons flooded areas of south Asia; rainfall 152 percent above normal flooded northeastern Australia; drought led to record-low water levels in the northwestern Amazon basin; sandstorms killed crops in the Middle East. Scientists confirmed the accelerated melting of the world's mountain glaciers as well as the great Arctic ice fields and ice sheets in Canada and Greenland. Satellites and ocean monitoring showed sea levels rising slowly because of glacial melting and the expanding thermal effect of warmer waters. The scorecard was not promising.

By the time negotiators arrived in Cancún on November 29, the deteriorating state of mankind's planet was no longer a distant risk debated by scientists. It was here and now and on every diplomat's lips. Scientists may shy away from blaming a specific event on global warming, but extreme weather and strange seasonal anomalies had traced a worrisome and now-familiar pattern over the face of the last decade. Heading into the Cancún negotiations, these events spoke volumes.

Huhne was blunt. "The pattern of these events and frequency of these events is due to climate change," he told the conference. He claimed the U.K. had paid out 4.5 billion pounds for flood damage in 2010 alone, compared with only 1.5 billion pounds in the previous ten years. This is why the U.K.'s Hadley Centre had combined with several other climate institutions to create what they call the Avoiding Dangerous Climate Change Program (AVOID). Advanced climate models showed that we probably have delayed too long to avoid serious damage from climate change. To give us a 90 percent chance of limiting temperature growth to below 2 degrees Celsius, AVOID's advanced climate models showed we would have to stop immediately all global emissions and we would probably have to employ "some geo-engineering intervention." For a 50 percent chance, we would have to reduce emissions at least 6 percent each year. Delaying action also means it will take longer to experience the benefits of emission reductions.

Scientists were getting panicky. A post-Copenhagen study by the Royal Society in the United Kingdom concluded that delays in reaching a global agreement to reduce greenhouse gas emissions were such that there is "virtually no chance of limiting warming to 2 degrees Celsius above pre-industrial temperatures."[17] For these scientists, the window of opportunity had closed. Current reduction commitments make it likely we will reach 3 or 4 degrees Celsius by 2100. Increases of this magnitude would wipe out farming in sub-Saharan Africa; raise sea levels between 0.5 and 2.0 meters, putting millions of people at risk; reduce water levels in all river basins in Africa, India, the eastern United States and southern Europe, most of which already experience some drop in levels;[18] and kill off coral reefs (the incubators of marine life) and large parts of the Amazon forest. The time available for adaptation will be reduced and the cost will be much higher than the US$100-billion annual Green Fund suggested in the Copenhagen Accord.

Meanwhile, time was running out on the world's only international treaty to reduce carbon emissions. The Kyoto Protocol's first commitment period was due to expire at the end of 2012 and support among industrialized nations for a second commitment period was weak. When I caught up to the now-former UNFCCC executive secretary Yvo de Boer at an energy conference in Montréal three months before Cancún, he said that given the present pace of negotiations, it could take ten years to secure a new global climate treaty.

Approaching Cancún, the annual rollout of scientific data didn't look good. And the political scene was equally disturbing. The climate change ledger was weighted heavily in favor of those who for decades had preached business as usual. Fools ruled.

With the pace of climate change accelerating, one would expect the politicians to sally forth to the Mayan Riviera with a new sense of urgency. But that's not what happened. Instead, they retreated into a luxurious Mexican bunker surrounded by thick lines of security. Gunboats patrolled the ocean and Mexican *federales* and the army manned roadblocks along the main highways leading into Cancún and the Moon Palace. Protesters could not get anywhere near the resort. And the science was relegated to the status of a trade show housed in a massive warehouse several kilometers away from the main event. The delegates were playing the maracas while the world burned.

The Mexican government emphasized a step-by-step approach towards a global agreement. Expect no treaty to be signed in Cancún, the Mexican secretary of foreign affairs, Patricia Espinosa, cautioned. Success would come in the form of small victories that eventually—perhaps in Durban, South Africa, in 2011, or perhaps later (who knew?)—could be fashioned into a legally binding global commitment. There was a time when negotiations were about reducing carbon emissions to avoid catastrophic climate change. At Cancún, the ultimate goal was saving the multilateral process itself.

But even this ambition had to absorb a blow of seismic proportion when, at the opening of the conference, Japan dropped a bombshell. "Japan will not inscribe its target under the Kyoto

Protocol on any condition or under any circumstances," the country's chief negotiator, Akira Yamada, told the conference. Under no circumstances, in other words, would the world's third-largest economy commit to reducing its emissions after 2012 under the Kyoto Protocol.

This rang the death knell for the world's only legally binding global agreement to address the challenges of climate change. Japan had intimated as much before, but never in such categorical terms.

"We know that developing countries insist on this commitment period," Akira Yamada said. "But it's not a fair or effective way to tackle climate change globally. The Kyoto Protocol only covers 27 percent of global emissions of greenhouse gases. We need the participation of all parties. It is as if we are in a soccer stadium and Annex 1 [industrialized nations except the United States] countries are the players and everybody else including the United States are the spectators. Just watching. We want all these major emitters to go down to the playing field and play together."

Canada, Russia and Australia rushed to support Japan. Europe remained supportive of Kyoto but with conditions. "We are willing to go into a second commitment period, but without the major emitters then you have no solution for the climate change conundrum," EU negotiator Peter Wittoeck said.

Any renewal of Kyoto would have to bind everyone, but particularly the United States, China and India, to some kind of emission reductions. The new reality was that China was now the largest carbon emitter on the planet and India's emissions were the world's fastest growing. The uncontrolled emissions of these countries would inevitably cancel out the reductions of industrialized countries, making the entire process counterproductive.

At any other conference, Japan's announcement would have driven developed and developing countries into two separate camps who would then pound each other with accusations of bad faith. But at Cancún this didn't happen. The reason was simple: They were not there to make momentous decisions. They were there to highlight areas of common ground and leave the rest to another day. A decision on the Kyoto Protocol and its post-2012 emission reduction pledges could wait until next year. They were there to keep the process rolling.

Every major climate change conference has a key word or phrase. In Copenhagen it was "ambitious." In Cancún the word was "balance."

Rich and poor countries acknowledged the importance of "balance." When U.S. chief negotiator Todd Stern arrived in Cancún, he used the word five times in this context in a brief opening statement to the media. "The key to an agreement here is that we get balance across the issues. I think there is an agreement to be had. I am quite sure of that actually. I'm not sure whether we will actually get it. That question hangs in the balance. But it is easy enough to see what the outlines of an agreement would look like and on the subject of balance, I think it is critical that there be genuine balance . . . If we can get that balance across all of the key issues we will unlock the door to an agreement."

The EU used it three times: "What we need is a balanced package of decisions, a package of decisions that will be balanced within these two negotiating tracks under the convention and under the Kyoto Protocol and a package that is balanced within each of these negotiating tracks covering all the elements of the Bali road map that was set out in 2007."

Initially, balance meant that everybody has to play a role in combating climate change no matter how rich or poor. It meant assuring that cuts in one country are not canceled out by increases in another. It meant that every country must pledge to reduce emissions in a transparent manner that can be monitored and verified by international inspection to ensure trust. Later, however, balance took on a different meaning. It meant everybody

gets what they want without having to commit to anything substantial—at least for now.

"Balance" was the brainchild of the United States. It suited its pur-
poses. America was still little more than a spectator in this process.
Its commitment in the Copenhagen Accord of a 17-percent reduction by 2020 was not legally binding. It had failed to pass any
legislation to back up that pledge. Instead, it relied on the imposition of EPA regulations to force reductions. Stern promised
Cancún that President Obama was looking at a "combination of
regulations and legislation" to meet its Copenhagen Accord
target. "There is more than one way to skin a cat," he said.

But would this American way be acceptable to the rest of the
world? I found South Africa's chief negotiator, Alf Wills, typing a
report to his minister while sitting next to the hot tub in the South
African suite. He said nobody believes the U.S. regulatory path is
credible because a new government could reverse it. What's more,
he said, the "EPA regulations are not good enough because
Congress can easily cut the budget." Unfortunately, Wills said,
"The United States has nothing else to offer."

In the lead-up to Cancún, the Americans and Europeans held
a series of meetings to discuss a joint strategy. The discussions reveal
how meek the process had become. On January 27, 2010, U.S.
deputy national security adviser Michael Froman and U.S. negotiator Jonathan Pershing had met in Brussels with Connie Hedegaard
and several other senior officials at the European Commission. The
meeting is described in two U.S. State Department cables released
by Wikileaks. Pershing emphasized the importance of persuading countries to adopt the Copenhagen Accord because "there
is no plan B for negotiation of a different agreement."[19] He went
on to express doubt as to whether India and China would honor
the language of the accord regarding monitoring, reporting and

verification. They discussed a British desire to include loan guarantees as part of the fast-track financing. But Hedegaard said: "Thirty billion dollars had been promised; it cannot be lent." Pershing replied, "Donors have to balance the political need to provide real financing with the practical constraints of tight budgets." The fast-track financing was quickly melting away.

Then the conversation turned to Cancún. The parties plotted their strategy. The cables revealed the underlying competitive approach to climate change diplomacy that has defeated progress in the talks. Froman and Hedegaard agreed that the European Union would muzzle its criticism of the United States and the two would work closely together to devise a strategy to "build up the fledgling Copenhagen Accord."[20] Froman claimed Copenhagen was an example "where both sides misread each other's negotiating bottom lines." He stressed that the EU model of "one-upmanship" did not work on the U.S. administration and the two sides needed to coordinate their efforts to bring the rest of the world onside and to isolate uncooperative countries. Froman noted that the BASIC countries (Brazil, South Africa, India and China) worked closely together to "impede U.S./EU initiatives and [play] the U.S. and EU off against each other." He added that "the U.S. and EU need to learn from this coordination . . . and work much more closely and effectively together ourselves, to better handle third country obstructionism and avoid future train wrecks on climate." In return, Hedegaard assured Froman the EU "was muting its criticism of the U.S. to be constructive."

In preparation for Cancún, the EU and the United States agreed to work together to rally the G77, Mexico, the BASIC countries and the small island states to their side. Hedegaard noted that "unhelpful countries such as Venezuela or Bolivia" would have to be dealt with, and she added that the EU is a "big donor to these countries." Froman replied, "We will need to neutralize, co-opt or marginalize these and others such as Nicaragua, Cuba, Ecuador."

The discussion then turned to the main goal of Cancún. Hedegaard said, "We must have universal acknowledgment that the world cannot afford failure to reach a binding agreement," and all countries had to agree to commit to 2020 targets. She told Froman that "it is vital to show economic benefits and potential job creation from bilateral cooperation on climate and clean-energy technologies, to build public support for our efforts." She promised to provide him with EU studies on the economic benefits of going green.

Froman said that the threat of carbon border taxes could bring China onside. He said the key to success at Cancún was to "avoid a damaging replay of our division."

The EU and the U.S.A. were strange bedfellows, worlds apart in their commitment to carbon reductions and clean energy. Europe had always been a leader, the United States a facilitator one moment and a troublemaker the next. But the results of their friendship pact were clear. At Cancún, the United States was no longer an issue. Nor were China and India and the rest of the G77 unrelenting critics of the process and the poor climate records of the industrialized countries. Stern had spent a day and a half in China in October with environment minister Xie Zhenhua in what he called "very intensive and good conversations." He also met several times with Xie outside China, and the two had a video meeting one week prior to Cancún. These negotiations resulted in China and India agreeing to a package deal that included some form of emission reduction targets and a truncated form of international monitoring, reporting and verification. But when it came down to putting real numbers on the table, none of these countries wanted to participate. Leave that for next year.

With India and China onside, the rest of the G77 followed. Many of the countries that had vigorously opposed the Copenhagen Accord, including Sudan, Venezuela, Tuvalu, Ecuador and Nicaragua, fell silent.

Bolivia, the strongest advocate of a 1.5-Celsius target, found itself isolated. Before Cancún, the United States cut off funding to that country and Bolivia expelled the U.S. ambassador. President Evo Morales came to Cancún preaching his message of Mother Earth as a deity and capitalism and globalization as her defilers. He alleged that the United States and European countries had bribed the other countries by threatening to cut off their financial aid. It wasn't the first time developing countries had complained about bribery. The response from Norway's environment minister had been to issue another threat: "If you want to accuse us of bribery, then we can eliminate any cause for accusation of bribery by eliminating the money."

American and European delegates quietly spread the word that the Bolivian leader was motivated solely by a desire to enhance his own and his country's global stature. Several European delegates volunteered that back home in Bolivia Morales was handing out mining leases like candy and without environmental restraints. Prior to his Cancún visit he had been in Japan, where he was negotiating financing to build an open-pit lithium mine and highly toxic processing plants.[21]

Behind the rhetoric of Morales's message was the fact that climate change was hitting Bolivia hard. Even the U.S. embassy in La Paz conceded that Bolivia is "already suffering real damage from the effects of global warming."[22] Its water resources were dwindling and mountain snow cover was disappearing. Morales saw no point in agreeing to a treaty that would allow the global mean temperature to rise any more than 1.5 degrees Celsius. Higher would mean certain death for his country and many others as well, he claimed, adding that the most recent science supported his position.

But in Mexico, Morales found no supporters. The major players stuck to 2 degrees Celsius. If the process were to prove legitimate, then everybody would have to rally around the number 2. Bolivia remained isolated.

In the delegates' zeal to show the world that the United Nations process worked, they left the two most fundamental issues of the negotiations—emission reduction targets and the legal framework or frameworks that would anchor these commitments—undecided.

Developing countries were still adamant that the Kyoto Protocol be retained for the simple reason that it was the only existing international agreement on climate change; years might pass before any replacement became legally binding. But with Japan, Russia, Canada and Australia refusing to participate in a second commitment period and the EU wavering in favor of one treaty, the two sides were stalemated. The solution was to create a third document that would link the Kyoto Protocol and the parallel treaty on Long-term Cooperative Action. This document would act as a common receptacle for emission reduction pledges. The actual emission pledges would ultimately be worked out at the 2011 conference in Durban, South Africa—or later.

Climate negotiations are a spiderweb of interwoven threads where one pull sends the whole thing quivering. During the last few days at Cancún, delegates split up into five ministerial working groups on the main issues of mitigation, adaptation, finance, forest preservation and carbon markets. Each country disclosed areas of compromise and areas where it would hold firm. The outcome was separate LTCA and Kyoto Protocol documents essentially containing something for everybody but commitments from nobody.

The agreements simply supplied the basis for more negotiations. The document on the renewal of the Kyoto Protocol was a mere two pages in which the parties agreed to complete the work of drawing up a post-2012 Kyoto agreement "as soon as possible." The agreement also "urges [parties] to raise the level of ambition of the emission reductions to be achieved by them." Linked to this agreement was the actual negotiating text, which still numbered seventy-two pages of options. In other words, it had not changed much since Copenhagen. The parties couldn't even decide on the format for the table showing each industrialized country's commitments. Nor could they decide the level of emission reductions or the length of the next commitment period.

The Cancún agreement on Long-term Cooperative Action was more substantial. The parties agreed to try to keep the global mean temperature increase below 2 degrees Celsius. This marked the first time this limit had been agreed upon in a United Nations document. But they did not lay out how it would be accomplished. The 29-page document also included verification and reporting requirements for emission cuts in developing countries as well as the US$30-billion fast-start financing scheme for 2010 to 2012 to help poor countries adapt to climate change. It created a US$100-billion annual Green Fund to begin in 2020, also to help poor countries adapt and reduce their emissions. It set up research centers to help the transfer of clean-energy technologies to poor countries. And it outlined the principles for forestry management and the preservation of carbon sinks as well as compensation for countries that preserve their rain forests.

This all sounds good until you realize that the LTCA document was designed only to provide a basis for further negotiations. When it came to real cuts in emissions, the document "emphasized" the need for developing countries to take the lead in making "deep cuts in global greenhouse gas emissions," but it didn't oblige them to do it. Nor did it define "deep cuts." In fact, it affirmed that the overriding priority for developing countries was economic growth. The document even acknowledged that their emissions would grow as they created jobs to eradicate poverty. "Social and economic development and poverty eradication are the first and overriding priorities of developing country Parties, and . . . the share of global emissions originating in developing countries will grow to meet their social and development needs." In other words, the agreement gave China, India, Brazil and any other country determined to grow its economy a blank check to spew whatever emissions it deemed necessary for economic growth and poverty eradication. When you consider that the poverty in countries such as China and India is largely of their own making, it is difficult to understand why they would get a free ride. Their poverty is intimately linked to overpopulation and poor governance, for which these countries alone are responsible.

The two agreements essentially kept the multilateral process going, one step at a time. Tragically, however, these steps were being taken on a treadmill. The parties simply agreed on two draft negotiating texts, hoping that further negotiations leading up to and at the South African conference in December 2011 would result in legally binding treaties. As the first paragraph of the LTCA agreement states: "Nothing in this decision shall prejudge prospects for, or the content of, a legally binding outcome in the future." In other words, everything was still up for grabs. Yet for public consumption, the Cancún decisions were tagged by delegates as a major breakthrough.

I knew that an agreement was close on the morning of the final day when I met up with Sudan's Lumumba. I asked him if he still believed that no agreement was better than agreeing to the two-degree benchmark. He told me my question was wrong. The right question is why are delegates ready to vote for an agreement to limit mean temperature increases to 2 degrees Celsius when they know that millions of people in more than 100 countries will be adversely affected. Why is it that "the meek always have to sacrifice or to be the sacrificial lamb?" But then he went on to say that he supported an agreement that would set 2 degrees Celsius today but provide for a review in 2015 for a 1.5 Celsius ceiling. A year ago he would never have agreed to this incremental approach. Now, on the last day of Cancún, he was onside. After the final vote he told me why.

"There are three reasons. First of all, on the procedural side, Mexico has done a great job. I mean resuscitating multilateralism, really doing what is right in terms of engagement of all etc., even if it is not perfect. They have done a great job and you have to give it to their political direction and true belief in multilateralism, their diplomatic ability and their ability to bring in multiple nations to start to really draft this. The second reason is very obvious, is that it is a provisional step forward. It is not legally binding. It is a basis for negotiating . . . The other reason why I do believe that it is an important document from my perspective, for the first time there is a very clear reference to the fundamental right of safeguarding human life, and that is the reference to Resolution 10–4 of the Human Rights Council. As you remember, in my thinking these negotiations are about two things: safeguarding life of humanity and equitable sharing of atmospheric space and right to development."

Cancún managed to give the impression that it had fostered a democratic, all-inclusive, consensus-gathering process. After twenty years of climate change negotiations, that was considered progress.

Once again, the package was pulled together after a twelve-hour marathon meeting that included twelve major countries and representatives of various national groupings. Chaired by Ambassador Luis Alfonso De Alba, Mexico's special representative for climate change, the meeting hashed out the final documents based on what the five ministerial groups had concluded. The Cancún agreements basically ended up with something for everybody and sent each country home with its own loot bag. In that sense it was, as Stern later said, a "balanced package of decisions."

Before the final votes, conference president Patricia Espinosa called the agreements "truly a renewal of the spirit that can work in the political world to come together in the best interests of humanity."

Bolivia disagreed. It was the only one.

Its UN ambassador, Pablo Solón, said he refused to sign "a document that means a significant increase in the average temperature that will put more human lives in a situation close to death . . . Two degrees Celsius is not acceptable . . . For countries like Bolivia who have snow on their peaks, it would mean the loss of this water. We have already lost one-third of this snow. There's a 50 percent possibility of having irreversible climate change. It is essential we work towards a goal that island states would not be threatened and that none of the states would disappear." But he was drowned out in a flash flood of enthusiasm.

I watched as a Saudi Arabian delegate tried to persuade Colombia to come to Solón's support. But the Colombian delegate remained silent.

If there was any turning in favor of Bolivia, Grenada's ambassador, Dessima Williams, put an end to it when she warned against holding out for the perfect agreement. "The perfect should not be the enemy of the good. We can still say that we left Cancún with something that we can work with."

Pablo Solón was not looking for a perfect deal, however. He was looking for a deal that protected his country.

The agreement committed the United States to nothing. Stern said, "It is not perfect, nothing like this ever is. It did not do everything we wanted, but it is very good from our point of view. It is a big step forward. What we have is a text that, while not perfect, is certainly a good basis for moving forward."

China's head of delegation couldn't have been more succinct: "We're basically satisfied."

The main driver behind the Cancún agreements was the worry that failure to demonstrate progress would mark the end of the United Nations process. It was a misplaced concern. If failure to show progress was the grave marker of multilateralism, these talks should have been buried years ago.

The true yardstick for these negotiations is the amount of carbon gushing out of our collective smokestacks. Despite decades of nattering, we continue to increase these emissions at a record pace. Cancún did not offer any hope that this trend will be reversed. Quite the opposite. The conference seemed to strengthen it, even institutionalize it. In a world smothered in lies, Cancún bamboozled with the best.

By the time Patricia Espinosa's light tap of the gavel ends COP 16, the sun has risen over Cancún. I gaze at the happy faces of the delegates all congratulating themselves on their great triumph. I watch as several Australian delegates hover around Stern, smiling and laughing in that kind of nervous, self-conscious way that some people have when they're trying to be friendly with their boss. And I think that the politics of the possible has reached a new low.

I leave the hall and catch a bus that takes me out of the Moon Palace for good. I then catch a second bus back to my hotel and go to bed. I wake in the early afternoon, slip on my bathing suit and go to the beach.

A pretty young woman who is French but was a sort of mercenary delegate for a Pacific island country is stretched out on a plastic deck chair, a bottle of beer stuck in the sand beside her. I linger over her beauty for a few long seconds and then walk into the ocean for a quick swim. I play volleyball with several members of the World Wildlife Fund. That evening I join friends for a meal of sushi. Nobody wants to talk about the convention.

The next morning I fly back to Montréal, mixed in with a load of tanned Canadians. Before the plane lands, we pull out our heavy coats and sweaters.

THE TROUBLE WITH OUR BRAINS

IN WHICH WE CONTEMPLATE
THE HUMAN RELUCTANCE TO CHANGE AND
HOW TO SOLVE THE DILEMMA OF GLOBAL WARMING

STATEMENTS ON THE SCIENCE OF CLIMATE CHANGE BY REPUBLICAN Party candidates running in the U.S. midterm Senate races in 2010:[1]

"There isn't *any* real science to say we are altering the climate path of the earth."
—**REP. ROY BLUNT, Missouri, Republican**
(Oil and gas contribution US$293,000[2]; US$29,200 from Koch Industries)

"I don't think there's the scientific evidence to justify it."
—**MARCO RUBIO, Florida, Republican**
(Oil and gas contribution US$177,000; US$31,200 from Koch Industries)

"I have been trained to read scientific documents, and it's malarkey."
—**SEN. TOM COBURN, Oklahoma, Republican**
(Oil and gas contribution US$180,350; US$14,800 from Koch Industries)

Testimony before Rep. Edward J. Markey's Select Committee on Energy Independence and Global Warming, Washington, August 2010:

"What troubles me and what keeps me up at night and into the office early in the morning is that we may well be past that tipping point and there is nothing we can do right now to avoid at least a one meter rise in sea level by the end of this century."
—DR. ROBERT BINDSCHADLER,
a senior fellow at NASA's Goddard Space Flight Center and a glaciologist

"Sometime within the next decade we may cross that tipping point that puts us warmer than the temperature at which Greenland can survive. Greenland by itself, if it melts, raises the global sea level on average about 23 feet. The deepest water in New Orleans after the hurricane was about 20 feet . . . [I was asked] 'What is the single strongest piece of evidence that holds up your global warming theory?' And my reply was 'What is the single thread that is strongest in a rope?' . . . It is an interwoven rope of evidence that shows very strongly that we humans are changing the world in a way that will impact us in major ways . . . There is no thread that you can cut out of that conclusion because it is so strongly based."
—DR. RICHARD B. ALLEY,
a geologist at Pennsylvania State University

As I mentioned earlier, the months between Copenhagen and Cancún were the hottest on record. Their temperatures exceeded those of the global record-warm years of 1998 and 2005. There was, however, one difference. In 1998 and 2005, a powerful El Niño helped hike the temperatures; 2010 didn't need this extra kick.

Russia got the brunt of it. Temperatures around Moscow and in most of the country's central and western regions soared to

42 degrees Celsius, 12 degrees Celsius above normal. They stayed high for weeks. By mid-July, Russia began to burn. Wildfires swept through more than 750,000 hectares of forest, destroying villages and crops and at one point threatening Moscow. Eastern Siberia was also hit. In the greater scale of annual Russian burns, the number of hectares lost was not abnormal. In heavily forested northern countries such as Russia and Canada, several million hectares on average are consumed each year by fires, caused mostly by humans. But the fires that hit Russia in the summer of 2010 were unusual in their ferocity, their natural causes and the amount of burn in a relatively short period of time. When it was all over, Russia had lost 10 million hectares (four times the norm), or a quarter of its arable land, to drought and fire. About 40 percent of its grain harvest was gone. Russia reacted by banning grain exports, which pushed world wheat prices up 42 percent. They hit a two-year high before sliding back in September. Only a bumper crop in the United States stopped prices from going through the roof.

But what if the record temperatures had hit the U.S. breadbasket? Lester R. Brown, president of the Earth Policy Institute in the United States, posed that question in September 2010 at the World Energy Conference in Montréal. The conference is held every three years and Brown was one of its keynote speakers. He laid out for his audience of corporate, government and institutional energy experts a devastating scenario.

He noted that Russia produces in a good year about 100 million tons of grain. The United States has a much more intensive grain harvest and gets bigger yields of about 400 million tons each year. If the same temperature anomalies had hit the United States, the world would have lost 160 million tons of grain. The effect on the world's grain prices would have been enormous. "The rule of thumb for ecologists is for every 1 degree Celsius rise in temperature, we can expect a 10 percent decline in grain yields," Brown said.

High temperatures causing fire and drought are not the only heat-related threat to the world's grain supplies. All the major rivers of Asia depend on glacier melt from the Himalayas and the Plateau of Tibet to sustain their flows through the dry seasons. These rivers irrigate the world's largest grain-producing areas. "We forget that China is the world's number one producer of wheat; India is number two; the U.S. is number three," Brown said. "China and India totally dominate the world rice harvest. So what happens to those glaciers will have an enormous effect, a direct effect, on the wheat and rice harvests of Asia." Furthermore, the melting of the Arctic and Antarctic glaciers will eventually flood Asia's rice-producing deltas. Asia will need to import increasing amounts of food. China already imports 70 percent of its soybeans, or half the world's production.

"How much and how fast do we have to cut carbon levels if we want to save at least the larger glaciers in the Himalayas and on the Tibetan Plateau, the glaciers whose ice melts sustain the major rivers of Asia?" Brown figures that to be safe, we would have to cut 80 percent in the next decade, a figure shared by many scientists. To achieve this, "we are looking at something approaching wartime mobilization." He and many others like him find hope in the remarkable industrial transformation that allowed Allied powers during the Second World War to produce hundreds of thousands of tanks, guns, ships and airplanes almost overnight. If we can rally nations to do that, we can do anything. Or so goes the logic.

Science may not be able to supply us with absolutes, but it can help us manage risk. The necessity to mobilize world resources towards a solution to global warming is, for many scientists, self-evident. The risk profile that science presents, and that is buttressed by the irrefutable physical evidence we see in such phenomena as melting glaciers, renders it insane to do anything less. Luckily, the solution is simple: Stop burning fossil fuels. It is a remedy that stares us down every time we choose to take a car instead of walking or riding a bike, every time we build another coal-fired power plant or continue operating an existing one instead of choosing a clean-energy alternative.

We are also lucky in that we have the money to make this transition. According to the International Energy Agency, the world will probably invest about $20 trillion over the next twenty years in the energy sector. Normally, most of that would go into oil, coal and gas. If we continue with this business-as-usual scenario—and barring a total collapse of the world economies—it's likely emissions will rise by perhaps as much as 50 percent by the middle of the century. On the other hand, if we invest most of that money into the installation of wind, solar, geothermal, nuclear and tidal energy systems strategically placed to maintain a steady supply of clean energy over integrated national and international grids— smart grids, as they are called—the transformation Brown and others refer to is possible. At the same time, we must invest heavily in research and development, insulate buildings, and redesign transportation systems that will reduce our carbon output. This is not a no-risk path. Every economic forecast predicts that measures to reduce our carbon emissions will slow growth. Jobs will be lost. Others will be transferred. Society will have to curb its energy-wasting habits. Yet even in the unlikely event that the scientific predictions on climate change are wrong, we will at least have created a much cleaner and much nicer world in which all species can enjoy a healthy future. But if we take no action or not enough action, and the predictions turn out to be accurate, which is more than a 90 percent probability, the impact on humanity will be devastating.

The mobilization required to make such an energy transformation has been successful in places such as Denmark, Germany and Spain, where giant wind turbines dot the landscape, feeding clean energy into millions of homes and businesses. In many cases these wind farms are owned by farmers and small municipalities that now profit from their investments. Slowly but surely, the realization that a simple windmill can empower even the tiniest community is taking shape.

At the beginning of this book I mentioned that when the Conservatives formed the government in Canada in early 2006, they canceled federal programs designed to help a province such as Prince Edward Island transform its energy grid from 93 percent fossil fuels based to at least 30 percent renewables by 2016. The small community of Summerside, which owns its own electrical utility, was not thwarted. It wanted to do its part to reduce greenhouse gases and at the same time lower its dependence on imported fossil-fuel energy from New Brunswick. A grid that once relied on diesel power now supplies energy to its sixteen thousand inhabitants primarily from four 3-megawatt windmills imported from Denmark. The city calculates that wind supplies about 50 percent of its energy and has allowed it to reduce its annual greenhouse gas emissions by 9,843 tons. The $30-million funding came from gasoline taxes and federal-provincial building funds. At peak performance, the city sells excess energy back to New Brunswick, which brings in about $1 million a year in net revenue. Backup comes from another wind farm across the island and from New Brunswick. The city has also installed smart meters in homes that sense the availability of wind energy and can turn on some appliances when it's windy. The city is helping residents with financial incentives to purchase space heaters capable of storing surplus wind energy in ceramic bricks. The bricks retain heat for circulation into the home when needed.

Summerside is fortunate in being one of the five windiest cities in Canada. But to help the city achieve its dual goal of clean energy and energy independence, the province created the economic framework that made it a good business proposition.

Its inspiration came from Denmark and Germany, which years ago created a green-energy revolution through a simple stimulus program designed by politicians. These two countries are now the largest producers of wind turbines. For both countries, alternative energy has become a multi-billion-dollar business with enormous export potential. In Germany, wind and solar industries employ more than 200,000 people. The small city of Summerside may be on the front lines of clean-energy conversion in Canada, but all of its technology is foreign, with the hundreds of patents held by German and Danish companies making it difficult for Canadian companies to establish a foothold in the business. Canada has some of the best wind corridors in the world, but to take advantage of them it will likely have to buy foreign technology.

One of the most innovative and successful German companies is Enercon. Its managing director and sole owner is Aloys Wobben, fifty-eight, an electrical engineer who started the company in 1984 in his backyard. It is now Germany's largest wind turbine manufacturer and Wobben has amassed a personal fortune of US$3.5 billion. His company employs more than 6,000 people worldwide and has shipped more than 16,000 wind turbines. He is part of an industrial sector that now generates about $68 billion in sales a year, with an annual growth rate of about 30 percent, and employs more than 500,000 people. The total number of wind turbines installed by the end of 2009 created enough energy equivalence to meet the electrical demands of the world's seventh-largest economy, Italy. Latin America alone is expected to reach 46 gigawatts of wind capacity by 2025—enough to supply about 500 million people.

The day I stopped by Enercon's booth at the Montréal energy convention, the company had launched its *E-Ship 1* transport vessel on its maiden voyage from Germany to Ireland carrying nine wind turbines. The ship uses four "rotor sails" in the form of vertical spinning cylinders, which rise from its deck, to help power its propellers. Wind energy allows the ship to reduce its carbon footprint by up to 30 percent, the company claims. The vessel will transport Enercon's windmills all over the world. The company hopes the design will open new avenues for reducing bunker fuel consumption by ocean vessels.

Wobben runs the research side, leaving the business of selling turbines to others. Lately he has been tackling the critical problem of energy storage. His company is experimenting with hydrogen technology to store power at the base of a windmill for when the wind dies.

Enercon probably would never have got airborne without the simple economic formula of feed-in tariffs started by Germany in 2000. They guarantee clean-energy producers priority access to the electricity grid at a rate of return high enough to cover long-term capital costs and add a decent profit. This encourages banks to lend money and investors to invest. "We have had 3,000 megawatts [MW] of renewable energy installation coming onstream every year since 2001," Dr. Hermann Scheer, the German politician who championed the feed-in tariff laws, told me. "In seven years, 22,000 MW altogether. And we created a new industry. It has opened the power grid to everybody. Fifty percent of the owners of windmills are small farmers." Germany's windmills produce an average of 1.2 MW per mill. "The newest windmills are 6 MW. That means that if we were to replace these 20,000 windmills by windmills in the average of 6 MW, we could have seven times more production. That means practically 50 percent of [Germany's] power consumption. This shows that [clean energy] can be introduced much faster than most experts think." For countries such as Germany, Denmark, Spain and China, these feed-in tariffs have catapulted wind and solar into important players in their energy mix while building up a clean-energy industry.

As I write, there are only sixty-four jurisdictions around the world with some form of feed-in tariffs. Many are cities, states or provinces. Yet the numbers are rising and they demonstrate once again that governance is important in creating the market formula for an energy revolution.

Among Europe's more ambitious projects is the Desertec Concept. It was conceived by a consortium that includes the insurance company Munich Re, Deutsche Bank, Siemens, the Swiss electrical company ABB, and several other banks and solar energy companies. The idea is to build a massive solar energy platform around the Mediterranean that will feed all of Europe, North Africa and the Middle East with electrical power. The plan will integrate into a so-called "smart grid" the solar power of the African deserts and the wind power of northern Europe. The Algerians claim they have enough solar power potential in their desert to power the world economy. Desertec plans to start similar projects in the United States, Australia, India and China.

Contrast the European approach with Canada's. Canada is blessed with some of the strongest and most consistent winds in the world, much greater and more reliable than anything Germany can muster. Yet its move in this direction is tentative at best. A case in point is Québec, a province that gets more than 95 percent of its energy from hydro and claims to be a world leader in clean energy. The ecological cost, though, is high. Few major and mid-sized rivers in Québec run freely. An area about half the size of France has been flooded and whole ecosystems destroyed. The reservoirs themselves produce greenhouse gases from the rotting biomass. The turbines also produce GHGs as they churn the gaseous water. Some hydro reservoirs in tropical countries such as Brazil produce more greenhouse gases than a coal-fired power plant. Colder regions produce much less of the gases, but the amount is not insignificant.[3] Still, Québec wants more dams. Its monopoly electrical utility, Hydro Québec, has decided to spend $6.5 billion building four hydro dams on one of the province's last remaining free-flowing rivers—the Romaine River, which runs 476 kilometers from Labrador to where it empties into the St. Lawrence River. The project, which will produce 1,550 megawatts of energy (enough to power about 600,000 homes), will destroy the ecosystem of that river. Its raging waters, its hurtling falls, its woodland caribou, black bears, moose, salmon and trout, its spectacular mountain vistas and rocky cliffs and dense forested valleys will be by 2020 a lost world, rotting at the bottom of vast reservoirs measuring 279.2 square kilometers. But Hydro Québec doesn't agree with this interpretation. "The transformation of the Romaine River is going to create a modified environment," a Hydro Québec official told me. She added that the utility plans to spend up to $200 million to "upgrade the aquatic environment" so that local residents can "practice kayaking, canoeing and a lot of recreational activities." As if the river had never supplied such recreation. Québec hopes to sell the electricity to the United States, but no contract has been signed.

There are viable alternatives.

Québec has some of the world's best wind corridors. Wind energy blends in nicely with existing hydro because the energy flow can quickly be adjusted from minute to minute. When the wind dies, hydro fills the void, and vice versa. Wind energy also permits reservoirs to fill, allowing hydro to store more energy during the winter when there is peak energy use and reservoirs become seriously depleted. The best wind corridors are also precisely where Québec has its largest dams, in the northern area of James Bay. So the distribution network is already in place. The Romaine River region itself is home to grade-A winds that are calm only 2 percent of the year.[4] One company proposed building a wind farm in the area that would produce the same amount of energy as the dams at half the price and take up only 9 square kilometers. The effect on the environment would be negligible. Hydro Québec rejected the idea out of hand. It builds dams, not windmills.

Energy conservation is another alternative. Québec buildings are energy pigs. Many older buildings have no insulation. Studies show that a $6.5-billion energy conservation program would not only produce more jobs than the dam project, it would free up enough energy to make the four new dams unnecessary. Geothermal heating from biomass and solar energy would also help in the energy mix.

But Québec's energy utility, its engineering companies and its construction industry and unions feed off big hydro projects. They are the internal organs of a monster that requires constant nourishment. Together they represent a powerful lobby in the provincial capital. Alternative energy sources are largely foreign owned and have little influence with provincial legislators. Québec is also a province whose net debt of $129 billion is the highest in Canada. Quebecers hold the fifth-worst net-debt-to-GDP position in the Western industrialized world.[5] The province needs to sell electricity to the United States to help relieve this terrible debt burden. The province also hopes to claim millions, if not billions, of dollars in carbon credits for its clean-energy hydro should the world ever develop a truly international carbon trading system. Conservation does not produce this kind of foreign income. Québec builds dams because that's what it does; because that's how it can control an energy monopoly; because it doesn't have to rely on foreign companies and technology that drain capital out of the province; and because that's how it makes money.

As I said in my introduction, we still do not have the technology to match the wonders of fossil fuels. We don't have the technologies that will continue to permit our high-speed lifestyles. Electric cars still cannot replace the power and flexibility offered by fossil fuel energy. Until science can find solutions to energy storage— and current work indicates they are well on their way—changes to our lifestyles will have to be made in the form of reduced energy usage. Nevertheless, we have the technologies that will allow us to live comfortably, if differently, in a clean and secure world, free of the risks posed by runaway climate change. Business is ready and able to jump in and fulfill the mandate. What we lack is the political strength to move forward on a global scale, which is essential to a speedy energy revolution. While some European countries forge ahead with clean-energy policies, countries such as Canada, China, Australia and the United States cannot be allowed to build more tar sands projects or coal plants. The Arctic partners cannot be permitted to exploit the oil in their polar regions. In one stroke, their actions can negate the efforts of the rest of the world. This is why the international climate talks are so important. They bring the required continuity to the goal of a global energy transformation.

But as we have seen, two decades of talks have failed to reduce emissions. This has been strictly a political failure. Even Europe, despite its leadership role, has a split personality. Italy for years has been led by a climate-change denier, as has the Czech Republic. Poland is more interested in building an economy on coal than on renewables. And in Great Britain, which first rang the alarm about global warming and whose governments have supported strong action, popular opinion wavered after media channels were flooded with denier rhetoric. Only countries such as Denmark, whose dependence on foreign energy sources forced them down the road to clean energy, have seen the light. Germany wisely recognizes the dangers of climate change, sees a path to energy independence and seizes an economic opportunity.

American politicians and media, meanwhile, still wallow in the dead-end argument over the validity of the science in a debate that is corrupted by politicians wearing For Sale signs, a warped ideology, television ratings and fossil fuel interests. Canada's government, a sort of Libertarian-Conservative hybrid, has actively suppressed information about climate change and has saddled itself with the most dangerously irresponsible leader in its history when it comes to the challenge we face. Like Canada, Australia has purposely built its economic policies around fossil fuels, creating and empowering a coal-fired monster. Not a country to miss an opportunity to feel important, its leadership of the Umbrella Group, where it poses as a responsible player, is a fraud.

Trapped in the fog of political wars over man-made climate change and what action should be taken, it's little wonder the public is confused. After all, the actions—or lack of actions—of their own countries send out mixed signals. What are people to think? There are those who have read the science and believe governments must take urgent action. There are those who believe in the science but don't think the problem is pressing. There are those who ignore the issue altogether, and there are those who remain unconvinced or think global warming is a hoax. It's not a recipe for a successful mobilization of civil society.

To understand more fully how politics, corporate power and climate change have affected our ability to act, it is worth exploring why people are still so reluctant to believe the scientific consensus and, even when they do believe it, decline to take action. For that we have to check out the human brain. How does our brain react when confronted with a challenge like climate change? Breakthroughs in brain imagery and brain wave analysis technology have shed light on the mental processes people go through when confronted with change. Neurologists are now able to track the split-second blood-flow responses in brain activity.

The brain's role is to keep our systems running. To do this, its fundamental impulse is to keep us sailing on an even keel, always seeking a state of healthy equilibrium, or what is called homeostasis—freedom from disease. Confronted with problems that cause stress and anxiety, the brain automatically seeks solutions so that life can return to normal.

Most of our actions flow out of over-learned habitual responses. Our home and work patterns are so routine we barely think about them. We fly largely on autopilot. These routines originally are worked out in the frontal cortex, where we deal with new ideas, thoughts and sensations. Once they become proven, workable and, to one degree or another, pleasurable, they migrate into a deeper region of the brain called the basal ganglia. Here the routine is firmly stamped onto a memory chip, rendering those actions or thoughts effortless and more or less instinctive.

The more you repeat an activity, the greater the likelihood that it will become automatic. "It's easier to follow a routine," says Dr. Lesley Fellows, a neurologist and associate professor at McGill University's Neurological Institute. "That's why when you go to the ice cream store with thirty flavors, it's rare that you actually compare them all. That would be a whole lot of trouble." Fellows notes that the actual energy expelled in a routine action is no less than the energy used, for example, to compare the ice creams. The routine action just *seems* effortless. Scientists are only beginning to study this sensation of effort. It could be a kind of trick the brain plays on us to keep us following paths that are tried-and-true—in a word, safe.

Change sparks into action a number of systems in the orbital frontal cortex and the amygdala where we deal with fear. (Rats have big amygdalae, which is one reason they are such adaptive beasts.) The amygdala emits a flight-or-fight response and the frontal lobes work out what to do about it.

Confronted with even the smallest challenge to our familiar patterns, the brain quickly emits fear signals that can disrupt attempts at rational thought. Our impulse is to get things back to normal—return life as we know it to a state of equilibrium. Dealing with newness takes effort. Slashing out new paths through the jungle of brain tissue is a sweaty task, which is why most people prefer to cling to the well-established brain thoroughfares.

But while rewiring can be exhausting, it also can have an exhilarating finale. There are areas of the brain that like to discover new ideas. And when they do, they emit a rush of neurotransmitters that give us an adrenaline high. It's the excitement of discovery. The problem is that it takes determination, courage and focus to journey off-road. It is essentially a question of summoning up the energy to focus. Some scientists theorize that the simple act of focusing on a problem slows the electromagnetic transmitters in the brain circuitry, allowing a new idea to establish new circuitry. When that circuitry proves workable, the reward is pleasure.

The trouble is that most people don't give that part of the brain a chance. Change for most people is too anxiety-provoking. Fear dominates. Climate change poses a complexity of stressful challenges with which few people know how to deal. "I can do something local like taking a bicycle," Fellows says. "There now, I did something—and I can forget about that. Or I can simply disbelieve the science and stop worrying about it. That is just self-protection."

California psychiatrist Jeffrey Schwartz believes there is an additional reason why it is so difficult to persuade most people to follow an unfamiliar path. Their brains are wired in such a way that they don't know how to think any differently than they have become used to, he says. The effort required to think any other way strikes them as too great. An oil executive trained to maximize profits has a completely different mental road map than an environmentalist. They see the world differently. An accountant sees the world differently from a lawyer, a baker from a truck driver, a journalist from a politician, a Conservative from a Liberal. People live their own reality and they stick to it.

How, then, do you get people to change? No one has really figured that one out. Certainly, trying to persuade people to change or to think differently rarely works. Fellows says there appear to be differences in brain systems when we initiate something ourselves as compared with when we are told to change. When we work something out and interpret the result independently, we get the reward of pleasure. When somebody else tells us what to do, even if it is to our benefit, the fear (and distrust) areas of the brain begin to work against us, forcing us back to a familiar equilibrium.

"The individually initiated behaviors are going to be backed up by dopamine release [reward sensations] if you get what you wanted out of the behavior. The more you do that, the more likely that behavior is going to continue," Fellows says. "In terms of the parts of the brain that get involved, it is definitely not the same thing for me to tell you what to do [as] for you to do it yourself and experience the outcome."

Also, according to Fellows, the brain isn't great at dealing with problems or threats that are nebulous, such as the distant dangers posed by climate change. The brain prefers to concentrate on the immediate. "If you eat piece after piece of chocolate cake, you will learn that chocolate cake is good and that will get stamped into your basal ganglia . . . To know that it is going to be bad for your cholesterol, [which] in turn is increasing your risk of heart attack, is a very nebulous concept. It's pretty astonishing that anyone ever changes anything."

The great strength of the brain is that it is highly adaptive to new situations if we choose to make the effort to change. Often this happens only when change is forced upon us. "If you ask a person what it would be like to be paralyzed, they would say it would be terrible, they wouldn't want to live like that," Fellows says. "But then you ask people who are paralyzed what life is like and they find different kinds of meaning and they rate their quality of life as almost identical to people who have never been paralyzed."

That, however, doesn't fully answer where people's beliefs come from in the first place and why they cling to beliefs that are patently inaccurate or that are completely contrary to an overwhelming consensus of experts. Modern psychology helps provide the answer.

Dan Kahan, a professor at the Yale law school, has joined a lineup of psychologists and other researchers who are trying to figure this out. His interest originally targeted how jurors view evidence and then broadened to how the public interprets scientific proof. He and his colleagues concluded that people search out expert opinion that conforms to their worldview of how society should operate. The credibility of science has little or no value in this process. A scientific consensus means nothing—people will create their own consensus. Kahan's work helps explain why the deniers of manmade climate change have been so effective. It doesn't take an expert to persuade many people that the experts are wrong. The deniers need only toss out a few pseudo-experts and a crowd gathers.

This possibly explains why men such as Lord Christopher Monckton, a regular pep-talker on the denier circuit, have been so successful. Nobody cares that he's not a scientist. All they care about is that he confidently tells them what they want to hear. So when he was hired as the wrap-up speaker at the Heartland Institute's denier climate change conference in Chicago in 2010, he confidently and, according to his worldview, quite accurately stated: "We [the deniers] are the consensus now. In the end, the truth is the center of every lasting consensus. And the truth is what is quietly pursued, often against overwhelming odds and fearsome hostility in academia, in politics and in the media, by those in this room who have kept academic respectability and scientific integrity alive and I thank you. Should we be triumphalist [sic]? Should we crow? Should we rejoice? Hell yes!! Send the forces of darkness sniveling back to their noisome lairs." The small crowd of mostly elderly people gave him a standing ovation.

Kahan's study[6] indicates that people who believe strongly in hierarchy, authority and, oddly, individualism are predisposed not to believe the scientific consensus that human activity is causing dangerous changes in the earth's climate systems, because manmade climate change poses a threat to their vision of absolute freedom. It triggers fear signals that banish rational thought. People with more egalitarian and community-based worldviews tend to believe in the scientific consensus.

This raises the question of whether our cognitive biases make rational public debate impossible. Kahan says there is no doubt that steps have to be taken to try to quiet these biases. He believes that scientists must find a way to quell people's fears by not making climate change "needlessly threatening to the identities of one or another group of culturally diverse citizens."[7] Clarifying a global plan of action would be a good start. Nothing sparks fear faster than the unknown.

We are also faced with the problem of stupidly confident people. These are the ones who always believe they are the smartest guys in the room and there is no telling them anything. David Dunning of Cornell University has been studying this phenomenon for years. He initially set out to find why it is that a large majority of people in almost any work environment believe that their performance is above average, which is statistically impossible. In other words, "Why are people typically convinced that they are more capable than they, in fact, are?" He discovered that it is usually the poorest performers who have the most inflated views of their performance, often placing themselves miles ahead of patently better performers.

A lot of it, he says, has to do with the fact that stupidity tends to breed unbridled confidence. Their incompetence "deprives them of the skills needed to recognize their deficits." In other words, they are just too dumb to know how bad they are. Conversely, smarter people tend to have less confidence in their skills because they are smart enough to understand their weaknesses. "If you are confident, clearly you are not going to error-check," Dunning says. "You are not going to take the extra step to verify whether or not you made the right decision. You are not going to check with other people. You won't go through the steps that would potentially help you to recognize when you have made a mistake. So to the extent that you are overconfident, you are less likely to take the steps that would inform you that you are being overconfident. And one of the biggest steps that you can do is consult with another person just to see if they have come to the same conclusion or not. This isn't that people are being defensive. They just think everything is fine. And as a consequence they are left with an overly rosy view of how they are doing."

Dunning thinks that part of the problem is that North Americans tend to be raised believing they have an innate skill that is unchangeable. "North Americans tend to think of themselves as finished products—either you have it or you don't. And that is quite different from another view you can have about skill, which is that it is all about hard work, that no one intrinsically has high or low skill. It is all about how much work you do, how smart you are in developing what skill you have."

Dunning says that if you have an overly positive view of yourself as a sort of self-made genius, then criticism or any evidence that might counter your self-image becomes threatening. "They face the challenge by running away, withdrawing, or disputing the person or the situation that is giving the feedback." He says climate change deniers, for instance, desperately and repeatedly cling to small but often meaningless discrepancies in the science as proof that they are right. "People who are deniers don't know what a good answer really looks like. They don't know what science looks like. They don't know what the real rules of science are and what scientists look for in deciding whether a conclusion is an appropriate one versus not. So the fact that there is consensus is huge to a scientist, but if you are on the outside you don't realize that the way science works it is very hard to come to consensus. So when you have consensus, that's a big thing."

Dunning says that people who regard themselves as a work in progress—in other words, they are trained in critical thinking—are far more likely to respond positively to new ideas and criticism. "If you have this developmental view, you have an improvement orientation that's 'Oh good, good advice, that's what I need to work on. Fantastic.'"

Could nations be like individuals, overconfident and too arrogant to take advice, with a worldview that is hardwired into "the solemnity of the remorseless working of things" and heading towards tragedy? The climate negotiations certainly encapsulate the hardwired fears of every country and its citizens. Sent forth by their nations are armies of diplomats who are themselves hardwired in the narrow diplomatic traditions of defending their country's self-interest. They are servants of fear, arrogance and distrust. Remember the Russian head of the delegation who dismissed the science of climate change as irrelevant to the negotiations? What mattered was defending his country. Remember the American Jonathan Pershing or the Canadian Michael Martin and their traditional "politics of the possible"? They were trained to think only in this one-dimensional world.

   We refuse to acknowledge that the enemy is not the other countries gathered around the negotiating table. The enemy is climate change. And it is uncompromising. It cares nothing for the vested business interests and individual conceits we use to justify our inaction. It does not recognize Australia's need to ship coal or Canada's need to dig up its tar sands. It doesn't recognize the United States' hopelessly partisan domestic politics or its arrogant refusal to bind itself to international treaties. Nor does it recognize China's and India's ambitions to maintain wild economic growth and supply cars to their billions of citizens. Climate change recognizes but one thing: an excess of tiny molecules in the atmosphere called greenhouse gases that absorb heat and then bounce it back to earth, warming the planet and taking us well beyond a livable heat zone. Stop the excess and the threat will recede. The degree to which we reduce our emissions is the only measure of success.

Perhaps, then, scientists, engineers, economists, technicians, lawyers, bankers and business experts should be rescued from the sideshow events and assembled at the center of the talks. These talks would no longer be negotiations; they would be planning meetings. Together, these experts would chart how to accomplish on a global scale the required reductions in carbon emissions. They are, after all, trained in critical thinking.

Governments do this sort of expert planning all the time to create fertile ground upon which money grows. Denmark's and Germany's green economies sprang out of expert planning. Canada's tar sands paid off not because business poured billions into it but because governments financed decades of research to develop the technology needed to extract bitumen from sand and turn it into synthetic oil. When that was achieved, governments lured business with enough environmental exemptions, tax breaks, royalty holidays, loans and grants to make it worth their while. The government stimulated the business by creating the base for a successful energy market, and investors responded. The business soon walked by itself and investors cheered the miracle of free enterprise, once again ignoring the now invisible hand of government.

The mandate for this expert group would be to prepare a global action plan involving all countries. It would set emission reduction goals according to the science, create institutions where money is poured into technological development, and design the ways and means by which each country can make a fast global transition to a clean economy. (We might start by educating the billions of poverty-stricken people in the world, who represent a huge reservoir of untapped talent.)

The advantage of such a plan is that it would clarify the way forward for all citizens of the world and show what in fact is possible. Planning at this scale would go a long way towards quelling the fears which create the biases that lead to disbelief and inaction. It would supply answers instead of questions. It would bring us out of the shadows and into a world of thoughtful global action.

It would bring honesty and focus to a corrupt political process. It would light the way forward.

In April of 2011, a brutal onslaught of extreme weather events, many of them record-breaking, raged through large areas of North America. Heavy rainfall and snowmelt flooded hundreds of thousands of acres of farmland from the Great Lakes to the Gulf Coast in 6 meters of water.[8] In Canada, record flooding hit Manitoba and Quebec. Tornados, drought and wildfires also reached record levels in many southern states. There were 875 preliminary tornado reports tripling the previous record set in April 1974. The estimated cost in property damage from all of these extreme weather events was $US32 billion. America had never seen anything like it. National Oceanic and Atmospheric Administration scientists said global warming was an important contributing factor and warned of worse to come.[9]

Climate change is a revolutionary force poised to roll over our capitalist world with a furious vengeance. It was our slavish devotion to the corporate creed and its contempt for human frailty that created global warming and now the monster mocks us and everything we have come to stand for. The challenges it poses have so far confounded all our efforts to tame the beast. The political failure is monumental. But that is only because we continue to place our faith in a political system hijacked by corporations and run by liars and propagandists on all sides of the spectrum. It seems then a lame hope to imagine a world where these same politicians would surrender power and award legitimacy to scientific and technical experts and then be prepared to accept and implement a global plan that would lay the framework for a new world vision based on compassion and respect for nature. But all else has failed. Our brains may instinctively force us to seek shelter in the reassuring equilibrium of the world we have come to know and trust. This is a false refuge. Yet our brains also have an amazing ability to chart new courses. Isn't it time we used them?

EPILOGUE

NOWHERE DID WE SEE ECONOMIC HIGH-HANDEDNESS PLAY OUT with greater zeal than at the United Nations climate change conference in November 2011 in Durban, South Africa. Aside from the now almost mundane fact that Durban repeated the failures of Copenhagen and Cancun by again delaying global commitment on effective emission reductions, Canada brought an extra degree of calculated cynicism to the table when it became the first industrialized nation to pull out of the Kyoto Protocol. Its excuse was that it cannot meet its treaty obligations. It claimed that the cost of that failure would be in the billions of dollars in terms of the added burden of emission reduction commitments in the post-2012 commitment period. To boot, it blamed non-industrialized nations for not cutting back their emissions thereby making it difficult for Canada to sell emission reductions at home. All of this is true. But it is deceitful. Canada's failure is intentional and entirely of its own making. It was a government decision to allow the uncontrolled growth of the tar sands projects in northern Alberta. It was a political decision that has made the Canadian economy so dependent on oil that the country has become a sort of one-industry town. To oppose the unfettered growth of the tar sands has now become heresy and a threat to

Canada's entire economy.[1] To contest the construction of a pipe-line earns a federal cabinet rebuke for being "radical." And those foundations that give financial support to environmental activists earn a visit from representatives of the prime minister's office carrying the underlying threat to decertify the foundation indulg-ing in political partisan activity. Plans have been approved to double the tar sands output by 2020, thereby more than tripling greenhouse gas emissions.

From an environmental monitoring point of view, the his-tory of the tar sands has been one of deceit and chicanery. Since 1997, the scientific monitoring of the tar sands environment has been the sole responsibility of the Regional Aquatic Monitoring Program (RAMP), which is funded entirely by industry. There was no peer review of its published studies and its raw data was kept secret. A peer reviewed study by federal scientists of RAMP's five-year report from 1997 to 2001 found that its data collection methods and statistical analysis had no credibility and its conclu-sions were "unsupported."[2] The study offered recommendations to improve the monitoring of the aquatic environment. But the federal government shelved the report and kept it secret until 2009. So nothing was done. RAMP continued with its phony sci-entific studies that allowed it to conclude in 2008 that "there were no detectable regional changes in aquatic resources related to oil sands development." Finally, independent scientific studies published by the *Proceedings of the National Academy of Sciences* demonstrated in 2009 that this conclusion was wrong. The tar sands industry was emitting enormous amounts of harmful toxins annually into the aquatic systems of the Athabasca that were having a toxic effect on fish and other aquatic life.[3] The studies also raised the possibility that these emissions could be a danger to human health. The work of these independent scientists ultimately forced the Canadian and Alberta governments and the tar sands industry in 2012 to admit that their monitoring had been totally inadequate.

New protocols were put in place and $50 million of industry money was set aside to create what the Canadian and Alberta governments called a "world class" environmental assessment program. The fact that some scientists both inside and outside of government had complained for years about this lack of oversight leads to no other conclusion than it was intentional.

Why is this important in the greater scope of climate change geopolitics? The tar sands project is the largest and most environmentally devastating energy project ever launched by mankind. It swallows hundreds of billions of dollars in investment capital from pretty well every major energy company in the world. The vested interests are global. It costs about $8 billion to build a tar sands project. Up and running, projects such as Suncor's earn more than $5 billion a year. We charge not one cent of that for the destruction of the environment. Instead, the industry drapes it in a mantle of lies claiming the tar sands are "ethical" and "clean" to confound our better judgment and appease those dark concerns that lurk deep in the shadows of every thinking person's brain. The tar sands are the battle ground for the continuance of the old economy. It is a battle we will all ultimately lose simply because the onslaught of climate change is unrelenting.

Geologists have recently concluded that mankind's influence on climate has become so powerful that we have entered a new geological age. They call it the Anthropocene, the Age of Man. Over the next few years they hope to formalize this new era. It is one of the great ironies of climate change that the Holocene, which for the last about 11,500 years has given mankind a stable climate in which to thrive, also gave him the evolutionary space to become his own destroyer.

April 2012

ACKNOWLEDGMENTS

I WANT TO THANK my longtime editor, Diane Martin at Knopf Canada, who has always been such a warm and enthusiastic supporter of my work; Anne Collins at Random House for her brilliant editing as well as Amanda Betts and John Sweet. There are many scientists, economists and members of NGOs who briefed me and supplied me with background material including Professor Peter Victor, Professor James Ford, Chris Huhne, Sharon Smith, Steven Solomon, Professor Christopher Green, Professor Peter Brown, Mark Lutes, Professor John Stone, Jim Bruce, Tom Harris, Chris Skrebowski, Professor Wayne Pollard, Professor James Drummond, John Drexhage and many others who can't be named because their careers would be compromised. Finally, I want to thank my wife, Janet, for her questioning, her debating and her (mostly) insightful comments. And my daughter, Katharine, for sitting down and reading the whole thing through.

NOTES

INTRODUCTION **TO THE COUNTRY FAIR**

[1] *National Inventory Report, 1990–2008*, Part 1, p. 30.

[2] guardian.co.uk, July 6, 2010, 10.38 BST

[3] These include Harper adviser and political scientist Barry Cooper of the University of Calgary, whose Calgary-based Friends of Science is among the most aggressive climate change denier organizations in Canada.

[4] http://www.sourcewatch.org/index.php?title=Oklahoma_and_coal

CHAPTER 1 **THE MERRY-GO-ROUND**

[1] Less than a month after the Canadian government ratified Kyoto, the Kochs, who are among the most active campaigners against environmental controls in the United States, sold their interest in Foothills to Petro-Canada. The deal had obviously been in the works for a long time. The leases are now owned by Suncor, which has yet to develop the project because of runaway labor and equipment costs.

[2] Kathryn Harrison, University of British Columbia, *The Struggle of Ideas and Self-Interest: Canada's Ratification and Implementation of the Kyoto Protocol*, prepared for presentation at the Annual Meeting of

the International Studies Association, San Diego, California, March 22–26, 2006.

3 Polls taken during 2002 showed that the majority of Canadians supported Kyoto. Even in Alberta, at least 50 percent supported ratification. Critics often claimed that Canadian support was based on ignorance of the treaty and of the fact that most of Canada's energy came from fossil fuels. But there was no indication that increased knowledge reduced support.

4 National Energy Board, *Canadian Energy Overview 2007.*

5 Canadian Energy Research Institute, *Economic Impacts of the Petroleum Industry in Canada*, July 2009.

6 *National Inventory Report, 1990–2008; UNFCCC Summary of Greenhouse Gas Emissions for Canada*, p. 1.

7 Canada's per capita GHG production is about 22.6 metric tons, but there is a huge variation between the provinces. Québec, which gets 98 percent of its power from hydro, creates 12 metric tons per person while Alberta, whose energy comes from coal and which is home to the tar sands, produces 70 metric tons. This in itself creates resentment over where the burden of emission reductions will fall. Alberta wanted compensation for cutting its emissions while Québec wanted compensation for having already cut its emissions by building hydro dams.

8 May's parents were politically active Democrats in New Haven, Connecticut. When Clinton was at Yale, he used to come to visit. When the Mays moved to Canada and settled in Cape Breton Island, they kept in touch.

9 Reuters, December 13, 2003. In 2003, Bedritsky became president of the World Meteorological Organization and in 2009 he became chief presidential envoy on climate change for the Russian Federation.

CHAPTER 2 **DENMARK'S HUBRIS**

1 In addition to my interviews with Connie Hedegaard, I also interviewed Per Meilstrup on several occasions. He was not only a close

adviser to Hedegaard and Rasmussen but also author of a book on the Copenhagen summit published in Danish. He wrote a summary in English called "The Runaway Summit: The Background Story of the Danish Presidency of Cop15, the UN Climate Change Conference," published in the *Danish Foreign Policy Yearbook, 2010.*

2 *Overview of the Reported Greenhouse Gas Emissions, 2009,* Government of Canada, December 2010, p. 4.

CHAPTER 3 **TRAGEDY IN COPENHAGEN**

1 *New York Times,* March 19, 2010.

2 www.probeinternational.org/carbon_credits; http://cdm.unfccc.int/index.html

3 Canada's 2009 *National Inventory Report* states that by 2007 Canada's per capita emissions had risen 6 percent above 1990 levels and its GDP rose 60 percent from 1990. Its total GHG increase was 26.2 percent above 1990 levels and 33.8 percent above its Kyoto target (excluding forestry and land use). With forestry and land use emissions, Canada's GHG emissions rose 47 percent. Its population rose 20 percent. Australia's GHG emissions increased 31.4 percent over 1990 levels by 2008 (excluding land use and forestry). Its population grew 23 percent. GDP tripled to US$1 trillion, from US$314 billion in 1990. The enormous increase is partially due to Australia's world-leading coal exports to countries such as Japan and China. The United States' GHGs increased 6 percent to six trillion tons while its GDP rose 60 percent, according to its 2009 *National Inventory Report.* Population rose 23 percent.

4 Article in *The Gazette* (Montréal), March 28, 2010, p. A3.

CHAPTER 4 **THE FAILING GIANT**

1 Greenland has been covered by an ice sheet for more than two million years, although there is evidence that parts of southern

Greenland hosted a boreal forest sometime between 450,000 and 900,000 years ago.

2   The glacial cycles in the northern hemisphere began about 2.5 million years ago. Prior to that, the north was much warmer. It is thought that one cause of this cooling was the rise of the Himalayas and other mountain ranges in the north. Air currents cool as they rise over the mountains, intensifying the cooling effect of orbital forcing.

3   http://www.csr.utexas.edu/grace/education/activities/pdf/Speed_Of_GRACE.pdf

4   The difference between ice caps, ice fields and ice sheets is a question of size and topography. Ice sheets cover more than 50,000 square kilometers. The earth has two ice sheets: Greenland and Antarctica. Ice caps are smaller in area and are dome-shaped and not constricted by mountains. Both ice sheets and ice caps comprise broad, uninterrupted expanses of glacial ice. Ice fields are the size of ice caps but are ribbed with glacial valleys and often bordered by mountain ridges. In any case, all of them are glaciers because they are made of ice that flows. So simply referring to them as "glaciers" is also accurate.

5   Fritz was a nickname given to him during the Second World War because his surname was German.

6   Unlike Norway's, Russia's and Alaska's Arctic regions, the Canadian Arctic is a barren, cold and formidable desert of rock, gravel and ice. Aside from hunting forays, not even the Inuit ever inhabited the High Arctic. Cold War politics and Canada's need to establish sovereignty ignited in the 1950s an urgent cultural, military and scientific drive to establish a Canadian presence in the Arctic. Climate change has revived this political and scientific urgency. That anybody lives up there at all reflects a degree of commitment that is sometimes hard to understand. Canadians still have only the vaguest idea of the region, which was handed to Canada by the British, who held a questionable claim, in 1880, soon after Confederation.

7   Scientific teams also conduct annual verification work on land ice in

central Greenland and on the Arctic island of Svalbard, Norway. Three teams verify sea ice readings in the Canadian Arctic off the coast of Alert, at the top of Ellesmere Island, in the Fram Strait, north of Svalbard, and the northern section of the Baltic Sea.

[8] *www.nsf.gov/od/opp/budget.jsp* The Office of Polar Research has a complete rundown of its budgets and expenditures. Canada's Polar Continental Shelf Program does not reveal its budget on its website. At its fiftieth anniversary celebration in 2008, Denis St. Onge, emeritus scientist at the Geological Survey of Canada, delivered a presentation entitled "Polar Shelf or Sovereignty on the Cheap."

[9] Wohlleben emailed me an infrared photo of the entire archipelago. It showed an array of capillary-like fissures and fractures stretching over the entire sea ice north and west of the High Arctic islands, as if the area were the outer wall of a giant human heart. The truly disturbing part, she said, is that the photos are of winter conditions, when the ice is usually "really settled."

[10] One theory posited by Dutch scientists claims that this plague-driven reduction in population caused the abandonment of farms, which in turn led to the natural reforestation of Europe. The trees absorbed CO_2 from the atmosphere, which then cooled the planet.

[11] "Form and flow of the Devon Island Ice Cap, Canadian Arctic," *Journal of Geophysical Research*, J.A. Dowdeswell, T.J. Benham and M.R. Gorman, Scott Polar Research Institute, University of Cambridge, Cambridge, U.K.; D. Burgess and M.J. Sharp, Department of Earth and Atmospheric Sciences, University of Alberta, Edmonton, Alberta; published April 10, 2004, p. 7.

[12] Given the unpredictable weather patterns, Polar Shelf maintains fuel caches all over the Arctic, including on sea ice. Because the sea ice can move twenty kilometers or more in a day, Polar Shelf has homing devices on the drums. The drums are picked up at the end of every season.

CHAPTER 5 **THE AGASSIZ VENTURE**

1 Jaelyn J. Eberle and John E. Storer, "Northernmost record of Brontotheres, Axel Heiberg Island, Canada,." Paleomological Society, 1999, p. 1.

2 "The vegetation of Fosheim Peninsula is unusually diverse for this latitude, with 140 vascular plant species occurring in the Hot Weather Creek area alone. This contrasts with the vegetation in similar materials in the western and central Queen Elizabeth Islands, which generally have less than 35 vascular plant species." Mean July temperatures at Hot Weather Creek taken from 1951 to 1980 are almost twice as high as those taken at the Eureka station. They were between 2 and 8 degrees Celsius, with a mean of 5. By the late 1980s the mean had increased to 12.7. July highs reached 20 degrees Celsius and for most of the month were above 15. Sylvia A. Edlund, Geological Survey of Canada, Ming-ko Woo and Kathy L. Young, McMaster University, "Climate, Hydrology and Vegetation Patterns Hot Weather Creek, Ellesmere Island, Arctic Canada," *Nordic Hydrology* 21 (1990): 273–286. Since then, mean temperatures at Hot Weather Creek have steadily risen.

3 Similar fossil forests have been discovered since 1881 on Axel Heiberg Island and elsewhere on Ellesmere. Scientists come from all over the world to study them, occasionally getting into nasty turf wars as they cart away sections of fossilized tree stumps. But the two main mappers of these finds were Jack McMillan and his fellow geologist Bob Christie, both with the Geological Survey of Canada, an organization to which Canadians owe a great deal for its often heroic mapping of the Arctic, its rock formations and its resources in extremely challenging conditions.

4 The reason some greenhouse gases are more powerful than others has to do with the fact that they operate in different frequencies in the light spectrum. Methane is considered about twenty times more powerful than CO_2 even though its life expectancy in the atmosphere stretches only a few weeks or months. Its strength is not based

on its molecules having more muscle or being more persistent; rather, it is because the spectrum in which they operate has a lot less competition. It's a bit like being the only girl at the prom. One lonely methane molecule will attract lots of radiation in its frequency while a CO_2 molecule in its crowded room may not get hit on at all. And if it does, who will notice? Different molecules absorb and emit in different frequencies.

5 The Holocene climate maximum—the "warm period"—should not be confused with the so-called medieval warm period that affected Europe from about 900 to 1200 CE. Climate records indicate that this more recent warm period was localized primarily to Europe and did not reach even the temperatures we have today, never mind the temperature highs reached during the Holocene climate maximum. Deniers claim that the medieval warm period was a time when Greenland was green. Temperature records from ice coring show that Greenland hasn't been green for millions of years except for its southeast coast, which went through a warm period about half a million years ago. It regularly greens in summer with help from the Gulf Stream, but that's just the lichen and Arctic grasses. According to the saga of Erik the Red, he named the island Greenland "because men will desire much the more to go there if the land has a good name."

6 Historical temperatures can be deciphered from ice cores by examining, among other things, the ratio of heavy water molecules, which scientists refer to as $H_2^{18}O$ or simply "oxygen-18," to lighter water molecules, or $H_2^{16}O$—oxygen-16—in deep core samples. Because it is a heavier isotope, oxygen-18 does not evaporate as quickly as oxygen-16. In cooler temperatures, less oxygen-18 will get into the atmosphere and travel to the polar regions. In warmer times, much more of this isotope will be found in polar cores. So there is a correlation between temperature and the amount of oxygen-18 in ice. Annual snow accumulations can be distinguished in the cores going back thousands of years. The coring of the raised bogs in peat lands also uses oxygen-18 techniques to correlate with

temperatures. Input into these bogs is in the form of rain or snow and output is evaporation or transpiration (plants). So they contain in their sediments an archive of dry and wet climate changes that stretch back four thousand years or more, scientists say.

7 Vinther et al., "Holocene thinning of the Greenland ice sheet," *Nature*, August 18, 2009.

8 NEEM project, "The last interglacial and beyond: A northwest Greenland deep ice core drilling project." http://neem.nbi.ku.dk/neeminfo.pdf/

9 Ice is an excellent transmitter of radar, and if the signal is powerful enough it can penetrate kilometers into the glacier. Water, however, tends to kill radar, which is how scientists discovered lakes inside Antarctica's ice sheet.

10 The entire list of countries: Belgium, Canada, China, Denmark, France, Germany, Iceland, Japan, South Korea, the Netherlands, Sweden, Switzerland, the United Kingdom and the United States.

11 Ideally, scientists would like to test for carbon dioxide in the pore spaces of ice cores dating back 140,000 years so they can assess the levels of atmospheric CO_2 during these glacial and interglacial periods. The trouble is, CO_2 is soluble in water. Greenland ice layers at the bedrock tend to melt due to geothermal heat and the CO_2 is dissolved. Another problem is, if you have dust that contains any carbonate, the water will dissolve the carbonate and that will release CO_2. Also, microbes live in the ice. They use organic carbon as an energy source and this produces CO_2 in the presence of water, so you find elevated CO_2 that has nothing to do with what was in the atmosphere at the time.

CHAPTER 6 **THE AXIS OF MELT**

1 Canada's ruling Conservative Party has a history of climate change denial. At the Arctic Five meeting, the government handed out a media package in which the words "climate change" and "global warming" never appeared. Instead, the government used the term

"altering weather patterns" to account for the trend in decreasing sea ice and glaciers.

2   Translated by Linda Jakobson in *China Prepares for an Ice-Free Arctic*, SIPRI Insight on Peace and Security, No. 2010/2, March 2010, p. 6.

3   U.S. Moscow embassy cable to secretary of state, June 9, 2009, Wikileaks, 212098.

4   U.S. Copenhagen embassy cable to secretary of state, November 7, 2007, Wikileaks, 129049.

5   The Svalbard archipelago measures 61,680 square kilometers. Christian Schneeberger, "Glaciers and Climate Change: A Numerical Model Study," Swiss Federal Institute of Technology, Zurich, p. 23.

6   See Swedish researcher Linda Jakobson's excellent report *China Prepares for an Ice-Free Arctic*, SIPRI Insight on Peace and Security, No. 2010/2, March 2010.

7   According to *Arctic Pollution, 2009*, by the Arctic Council's Arctic Monitoring and Assessment Program: "There have also been discussions about the possibility that the stored waste would set off a spontaneous nuclear chain reaction. Current assessments show that, even though conditions are worse than expected, a chain reaction could not be initiated without an external influence. There remains uncertainty about what the consequences of such an accident would be for the nearby area and surrounding region.".

8   Telenor website: www.telenor.com/en/news-and-media/press-releases/2008/us-federal-court-grants-telenor-motion-holds-altimo-in-contempt-imposes-fines-and-orders-altimo-to-sell-shares

9   Anker, Morten, "The High North and Russo-Norwegian bilateral economic relations," p. 41, www.fni.no/russcasp/MA-bon1009-32-41.pdf

10 Linda Jakobson, *China Prepares for an Ice-Free Arctic*, SIPRI Insight on Peace and Security, No. 2010/2, March 2010.

11 *AAA-06 Arctic petroleum provinces (iii): Petroleum geoscience of the North American and Greenland basins, Part 2*. International Geological Congress Oslo, 2008. However, Professor Andrew Miall, a veteran Arctic and petroleum geologist and former head of Canada's

Academy of Science, dismisses the argument of the Lomonosov Ridge as a "stunt." He says the Lomonosov is a "fragment of continental crust" with only a thin sedimentary crust where there is little more than a "slight" chance of finding oil.

CHAPTER 7 **THE ARCTIC'S POISONOUS SECRET**

[1] According to documents obtained by Postmedia News through access to information laws, Natural Resources Canada instructed its scientists in the spring of 2010 that they had to obtain ministerial approval to speak to the press about high-profile issues such as "climate change, oil sands" or any subject in which a reporter from national or international media expresses interest.

[2] I later obtained satellite photos taken between August 7 and 29, 2010, showing the shelf quite suddenly breaking into thousands of pieces before the Arctic sea ice packs them together again.

[3] http://www.atsdr.cdc.gov/toxprofiles/phs46.html#bookmark04

[4] A. Steffen et al., "A synthesis of atmospheric mercury depletion event chemistry in the atmosphere and snow," *Journal of Atmospheric Chemistry and Physics*, 2008.

[5] Dr. Jan Bottenheim, who worked with Schroeder at Environment Canada on mercury depletion, discovered that the same chemical process each spring takes out the entire Arctic ozone layer. During our telephone interview, an Environment Canada flack named Tracy Lacroix came on the line and instructed me not to ask any questions about climate change, claiming Bottenheim knew nothing about it—as if climate change were some kind of segregated academic field. I ignored her, as, thankfully, did Bottenheim.

[6] A. Steffen et al., "A synthesis of atmospheric mercury depletion event chemistry in the atmosphere and snow," 2008.

[7] One gallon of jet fuel emits 21.1 pounds of carbon dioxide, according to U.S. Energy Information Administration, fuel emission factors; 6.76 pounds/gallon. Converted to metric tons.

CHAPTER 8 **WHAT OIL?**

1   Total exports of goods to the U.S.A. in 2009 were US$224.9 billon, of which oil and gas shipments totaled US$122 billion.

2   "Notification of Proposed Research Cruise, RV *Polarstern*," Alfred-Wegener-Institut für Polar-und Meeresforschung, Germany.

3   Natural Resources Canada Community Engagement Report, Grise Fiord Community Meeting, September 16, 2009.

4   Ibid.

5   Ibid.

CHAPTER 9 **PROPOSITION 23 AND THE MIRROR ON AMERICA**

1   "Climate Change Scoping Plan, A Framework for Change," December 2008, Government of California, p. 7.

2   The Howard Jarvis Taxpayers Association is a well-financed, angry dog. When it gets its teeth into something, it doesn't let go. So when the state's attorney general, Jerry Brown, who opposed Proposition 23, drafted the description of the petition for the ballot, the association sued him, claiming the title and summary were "false, misleading and unfair." The association convinced the judge to change the wording on the ballot that stated the proposition sought to "abandon" the climate change law. The association wanted "abandon" replaced with "suspend." It also demanded the word "polluters" be changed to "sources of emissions."

The original title description read: ABANDONS IMPLEMENTATION OF AIR POLLUTION CONTROL LAW (AB 32) REQUIRING MAJOR POLLUTORS TO REPORT AND REDUCE GREENHOUSE GAS EMISSIONS THAT CAUSE GLOBAL WARMING, UNTIL UNEMPLOYMENT DROPS TO 5.5 PERCENT OR LESS FOR FULL YEAR.

The judge changed it to: SUSPENDS IMPLEMENTATION OF AIR POLLUTION CONTROL LAW (AB 32) REQUIRING MAJOR SOURCES OF EMISSIONS TO REPORT AND REDUCE GREENHOUSE GAS EMISSIONS THAT CAUSE

GLOBAL WARMING, UNTIL UNEMPLOYMENT DROPS TO 5.5 PERCENT OR LESS
FOR FULL YEAR.

3 http://www.aqmd.gov/comply/1118/rpts/2010/TesoroWilm10.htm;
   http://www.aqmd.gov/comply/1118/rpts/2010/Ultramar10.htm

4 projects.publicintegrity.org/oil/report.aspx?aid=347

5 Koch Industries website: www.kochind.com/ViewPoint/lowCarbon.aspx

6 March 1, 2011.

7 www.opensecrets.org

8 Jane Mayer, *New Yorker*, August 30, 2010; and Greenpeace, "Koch
   Industries Secretly Funding the Climate Denial Machine," March
   2010, http://www.greenpeace.org/usa/en/media-center/reports/
   koch-industries-secretly-fund/.

9 www.heritage.org/about

10 http://mercatus.org/publication/environmental-protection-agencys-
   request-comment-petition-control-emissions-new-and-use-. See also
   the Greenpeace report entitled "Koch Industries Secretly Funding
   the Climate Denial Machine," March 2010.

11 Cal-access.ss.ca.gov

12 The National Academies include the National Academy of Sciences,
   the National Research Council, the National Academy of
   Engineering, and the Institute of Medicine.

13 According to Environment Canada, tar sands emissions increased 11
   percent in 2009 despite industry claims that they were falling.
   Canada's 2009 *National Inventory Report* on greenhouse gas emissions
   shows that while tar sands emissions continued to rise, Canada's
   overall emissions fell in 2009 about 6 percent largely because of the
   recession. Mark Johnson, a spokesperson for Environment Canada,
   told me in May 2011 that "oil sands emissions increased from 5.5
   percent of (Canada's) total in 2008 to about 6.5 percent in 2009."

14 In December 2009, U.S. ambassador David Jacobson sent a cable
   to Washington in which he stated that Québec premier Jean
   Charest might have been influenced by Power Corporation, the
   huge Canadian financial conglomerate, to soften his criticism of

Ottawa for its weak climate change policies when Charest attended the Copenhagen climate change conference. The cable, released by Wikileaks, stated that "Whether (Jean) Charest was influenced by Power Corp. to tone down his criticism of the federal government is unclear, but the corporation's provincial and federal influence is undeniable." The cable noted that Power Corp. is the major minority shareholder in the French oil company Total S.A., which has invested $6 billion in the tar sands. At the time, *La Presse*, the Montréal daily owned by Power Corp., criticized Charest in an editorial, calling his Copenhagen statements "arrogant" and "disloyal to Ottawa."

[15] http://www.ico2n.com/

[16] Fugitive Emissions Projections 2010, Australian Government, Department of Climate Change and Energy Efficiency.

[17] A metric ton of coal generates about 2.7 tons of CO_2; http://greenhouseneutralfoundation.org/articles/2010/05/24/king-coal-in-australia-the-ugly-political-truth/

[18] As Jessica Irvine from *The Sydney Morning Herald* calculated in her February 11, 2011, column: "For every dollar the mining lobby spent fighting the tax with emotive ads, featuring wholesome-looking miners, it saved another $2,750."

[19] *Proposed Carbon Pricing Scheme*, Seamus French, published by Minerals Council of Australia, 2011.

[20] http://cal-access.ss.ca.gov/Campaign/Committees/Detail.aspx?id=1323934&session=2009&view=received

CHAPTER 10 DEAD ZONES

[1] John J. Magnuson et al., "Historical Trends in Lake and River Ice Cover in the Northern Hemisphere," *Science* 289, September 8, 2000. The time series range from 1846 to 1995. Some records go back much further and indicate that changes to the length of the ice season were already occurring in the eighteenth century. There was

a reverse trend towards a longer ice season from about 1872 to 1897, but this is the single exception, the authors say.

2   Philipp Schneider and Simon J. Hook, "Space observations of inland water bodies show rapid surface warming since 1985," *Geophysical Research Letters* 37, 2010.

3   Jay A. Austin and Steven M. Colman, "Lake Superior Summer Water Temperatures," *Geophysical Research Letters* 34, 2007; Jay Austin and Steve Colman, "A Century of Temperature Variability in Lake Superior," American Society of Limnology and Oceanography, 2008.

4   Ankur R. Desai1, Jay A. Austin, Val Bennington and Galen A. McKinley, "Stronger winds over a large lake in response to weakening air-to-lake temperature gradient," *Nature Geoscience* 693, 2009.

5   "Confronting Climate Change in the Great Lakes Region: Impacts on Our Communities and Ecosystems," a report by the Union of Concerned Scientists and the Ecological Society of America.

6   Cindy Chu,* Nicholas E. Mandrak and Charles K. Minns, "Potential impacts of climate change on the distributions of several common and rare freshwater fishes in Canada," Fisheries and Oceans Canada.

7   Ibid.

8   "Threats to Water Availability in Canada," Environment Canada, 2004.

CHAPTER 11 THE TEMPTATIONS OF THE MOON PALACE

1   Compiled by DARA, a Madrid-based international organization that assesses the effectiveness of aid to poor and vulnerable countries, in partnership with the Climate Vulnerability Forum. See daraint.org.

2   I once asked Yvo de Boer, executive director of the UNFCCC, what will happen to the Netherlands when the seas rise. His reply was immediate and tinged with a sense of pride: "We're a wealthy enough nation to adapt."

3   This report combines datasets from the U.K.'s Hadley Center of the Met Office and the Climatic Research Unit, University of East

Anglia; the U.S. National Oceanic and Atmospheric Administration; and NASA's Goddard Institute of Space Studies. The report is peer reviewed.

4 The Climate Action Network Canada published a report in November 2010 outlining Canada's lobbying attempts to persuade foreign governments not to enact clean-fuel legislation. It was based on documents obtained through access to information legislation. See http://www.climateactionnetwork.ca/e/news/2010/release/index. php?WEBYEP_DI=66

5 Ambassador Jacobson cable to U.S. State Department, November 5, 2009, Wikileaks.

6 2009 Victoria Bushfires Royal Commission Final Report, http://www.royalcommission.vic.gov.au/Commission-Reports/Final-Report/Summary/Interactive-Version.

7 Italics inserted by author.

8 Australia National Greenhouse Gas Accounts, *National Inventory Report 2008*. In 1997, Australia negotiated a clause into the Kyoto Protocol that allowed it to include the halting of deforestation in its 1990 reference level. In other words, because it has not destroyed 108 million hectares of forested land, this carbon sink means that Australia can now claim to have increased its emissions only 9 percent. It's one of those clever loopholes countries become so proud of and climate change drives a cyclone through.

9 "Analysis of options to move beyond 20% greenhouse gas emission reductions and assessing the risk of carbon leakage," European Parliament Communication, Brussels, 26.5.2010, COM(2010) 265 final.

10 Ibid., p. 5.

11 Ibid., p. 8.

12 http://www.voxeu.org/index.php?q=node/3156

13 *World Energy Outlook 2009*.

14 "Free allocation of emission allowances and CDM/JI credits within the EU ETS," Oko-Institut for Applied Ecology. www.oeko.de and www.wwf.de

[15] Earth System Research Laboratory, Mauna Loa, Hawaii, National Oceanic and Atmospheric Administration, United States. www.esrl.noaa.gov/gmd/ccgg/trends/#mlo

[16] As per data collected by NASA, NOAA and the UK's Met Office among other meteorological institutions.

[17] Mark New, Diana Liverman, Heike Schroder and Kevin Anderson, "Four degrees and beyond: the potential for a global temperature increase of four degrees and its implications," Royal Society Publishing, 2010.

[18] Fai Fungi, Anna Lobez and Mark New, "Water availability in +2°C and +4°C worlds," *Journal of the Royal Society*, November 29, 2010.

[19] February 10, 2010, cable from Brussels to U.S. secretary of state, Wikileaks.

[20] February 10, 2010, cable from Brussels to U.S. secretary of state, Wikileaks.

[21] Bolivia has the world's largest lithium resources, but they have yet to be mined because they are in a remote area of the country that lacks water and infrastructure and also because Bolivia demands 60 percent of the profits.

[22] U.S. embassy La Paz cable, February 2010, Wikileaks.

CONCLUSION **THE TROUBLE WITH OUR BRAINS**

[1] Compiled by the Center for American Progress.

[2] http://www.opensecrets.org/politicians/summary.php?cid=n00005195

[3] Alain Tremblay, Louis Varfalvy, Charlotte Roehm and Michelle Garneau, "The Issue of Greenhouse Gases From Hydroelectric Reservoirs: from Boreal to Tropical Regions." Springer, 2007. Emissions from reservoirs decline after about ten years. These so-called fugitive emissions are often not reported in national GHG inventories.

[4] http://atlas.nrcan.gc.ca/site/english/maps/archives/3rdedition/environment/climate/020

[5]   www.conferenceboard.ca/topics/economics/budgets/
      quebec_2010_budget_EN.aspx
[6]   D. Kahan, D. Braman and H. Jenkins-Smith, *Cultural Cognition of
      Scientific Consensus. Journal of Risk Research*, DOI: 2010.
[7]   Ibid.
[8]   *Spring 2011 U.S. Climate Extremes*, National Oceanic and
      Atmospheric Administration, National Climate Data Center: www.
      ncdc.noaa.gov/special-reports/2011-spring-extremes
[9]   *New York Times*, John M. Broder, June 15, 2011; Associated Press,
      Randolph E. Schmid, June 16, 2011.

EPILOGUE

[1]   In his Jan. 13, 2012, column in the *Times Colonist*, Postmedia
      pundit Michael Den Tandt claimed that "Canada's future economic
      prosperity depends" on the proposed Northern Gateway Pipeline,
      which is designed to bring tar sands oil to the British Columbia port
      of Kitimat for shipment to Asia and California.
[2]   *(RAMP) Scientific Peer Review of the Five Year Report (1997–2001).*
      (Fisheries and Oceans Canada, Winnipeg, Manitoba, Canada).
      Page iv. "The problems with the report are found in lack of details
      of methods, failure to describe rationales for program changes,
      examples of inappropriate statistical analysis, and unsupported
      conclusions. That being said, the reviewers raised significant concerns
      about the Program itself. They felt there was a serious problem
      related to scientific leadership, that individual components of the
      plan seemed to be designed, operated and analyzed independent of
      other components, that there was no overall regional plan, that clear
      questions were not been addressed in the monitoring and that there
      were significant shortfalls with respect to statistical design of the
      individual components."
[3]   "Oil sands development contributes polycyclic aromatic compounds
      to the Athabasca River and its tributaries", Erin N. Kelly et al,
      Proceedings of the National Academy of Sciences, December 29, 2009.

INDEX

WILLIAM MARSDEN is author of the National Business Book Award–winner *Stupid to the Last Drop: How Alberta Is Bringing Environmental Armageddon to Canada (and Doesn't Seem to Care)*, and co-author with Julian Sher of the national bestsellers *Angels of Death: Inside the Bikers' Empire of Crime* and *The Road to Hell: How the Biker Gangs Are Conquering Canada*. He is a senior investigative reporter for the *Gazette* in Montréal. wmarsden@bell.net